Valuing Wind Generation on Integrated Power Systems

Valuing Wind Generation on Integrated Power Systems

Ken Dragoon

Thanks for buying my book!
Ken Dragoon
4/25/11

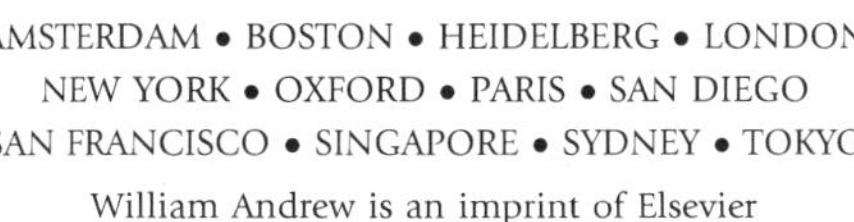
AMSTERDAM • BOSTON • HEIDELBERG • LONDON
NEW YORK • OXFORD • PARIS • SAN DIEGO
SAN FRANCISCO • SINGAPORE • SYDNEY • TOKYO
William Andrew is an imprint of Elsevier

William Andrew is an imprint of Elsevier
The Boulevard, Langford Lane, Kidlington, Oxford OX5 1GB, UK
30 Corporate Drive, Suite 400, Burlington, MA 01803, USA

First edition 2010

British Library Cataloguing in Publication Data
A catalogue record for this book is available from the British Library

Library of Congress Cataloging-in-Publication Data
A catalog record for this book is availabe from the Library of Congress

ISBN–13: 978-0-8155-2047-4

For information on all William Andrew publications
visit our web site at books.elsevier.com

Printed and bound in the USA

10 11 12 13 14   10 9 8 7 6 5 4 3 2 1

Contents

Preface

Thank you for sending me a copy of your book—I'll waste no time reading it.

Moses Hadas, 1900–1966

The purpose of this book is to help power system analysts, consultants, and regulators understand, undertake or better understand wind integration studies. A significant diversity exists in the power industry with respect to operating practices, level of sophistication in the industry's analyses, and in the very language used. For that reason, there is no single way to serve all analytical needs with respect to wind valuation studies. A range of approaches to the analytical problems will be presented where possible. A fairly complete glossary is included in the back of the book to address the various usages of terms such as 'capacity' and 'reserves'. It is not practical at this point to suggest that common definitions of certain terms be adopted. The purpose of the glossary is to make as clear as possible what is meant by important words and terms as they are used in this book.

Acknowledgements

There is no way to adequately express the gratitude I feel toward everyone who contributed to this book. First of all must be Phil Carmical, who suggested the seemingly crazy idea that I might write a book about wind power. I am also totally indebted to Rachel Shimshak and the Renewable Northwest Project who gave me my all-time dream job, and who encouraged this work.

I have learned the hard way that books are really the work of many people who have to set the author straight. It is hard to find the words to thank all the contributors who tried to keep me on the right track, helping make the book reasonably complete and pointing out where I may have gotten things wrong or just garbled. The remaining errors are all mine (one of the perks of writing a book). This work would have been much less complete, accurate, and understandable without the generous help of subject matter experts: Brendan Kirby, Michael Milligan, Hannele Holttinen, Charlie Smith, Cameron Potter, Justin Sharp, and Esben Hegnsholt Olsen. Special recognition is due to Brendan Kirby and Charlie Smith, who came through when time was short. Thanks also to Michael Schilmoeller, Katie Kalinowski, Sam Lowry, Diane Nowicki, and Cindy Towle for their editing assistance.

Thanks also to Steve Wasserman for acting on my behalf in a first-ever book endeavor. Thanks to my wife and family for putting up with my being missing in action nights and weekends to put this work together. And, of course, thanks to all the readers of this book who will help move wind power into the mainstream as a resource forming a vitally important part of the generating resource mix in the USA and abroad.

Chapter | one

Introduction

Wisdom sails with wind and time.

John Florio, linguist and lexicographer, 1553–1625

Wind power has become one of the fastest growing technologies in the world for producing electric power (WWEA, 2009). Environmental concerns are driving public and private industry policies that help promote this relatively economical, emission-free source of electric power. The US Department of Energy reports that wind energy may supply 20% of all electric power consumed in the USA by the year 2030 (US DOE, 2008), the percentage roughly equivalent to the load currently served by nuclear energy. Although wind energy production in the USA recently topped other nations, Europe still leads North America in installed wind generation by nearly two to one (WWEA, 2009), planning to reach 30% or more. The top four countries in percentage of energy generated by wind generation are Denmark (20%), Portugal (15%), Spain (14%), and Germany (9%). Figure 1.1 shows the growth of installed wind generating capacity from 2001 to 2010.

Several factors fuel European dominance over the USA. Chief among them is stronger government policies in Europe mandating renewable energy development as an important means to address the climate change crisis. Other factors may also be at work. Historically tasked with finding the most cost-effective means of ensuring reliability of power systems, American planners and power system operators are not natural allies of a resource perceived to be more costly and less reliable than traditional power sources. The prospect of meeting load with a resource as variable and uncertain as wind power is often accompanied by a rather skeptical attitude over how such a power source can reliably and economically meet demand. This book is intended to help analysts understand the value of wind energy in the context of complex power systems and to demonstrate that wind energy can contribute to meeting load in an economical, efficient, and reliable manner.

Valuing Wind Generation on Integrated Power Systems. DOI: 10.1016/B978-0-8155-2047-4.10001-8

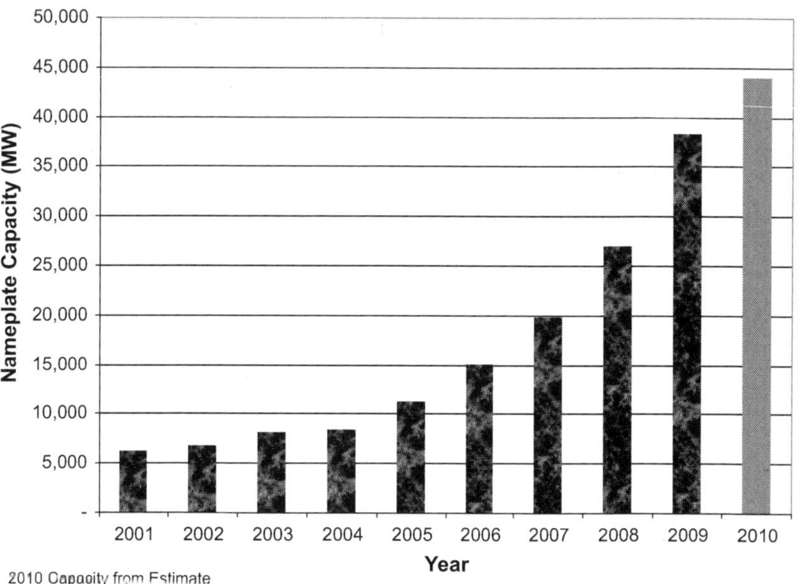

FIGURE 1.1 Installed worldwide wind-generating capacity. *Source: World Wind Energy Report 2009.*

Every type of electric power generation brings its own peculiar set of advantages and disadvantages. Coal and nuclear plants tend to generate at relatively consistent levels, but are subject to significant maintenance outages for repair work or refueling. They have limited ability to adjust to the dynamic nature of demand, and may take days to reach full output from a full shutdown. Power plants fueled by natural gas tend to be relatively flexible in changing output to meet the dynamic characteristics of demand through the day, but are also subject to relatively volatile fuel prices. Hydro units are often some of the most flexible generation resources in a utility portfolio, but have complex time-dependent behavior that can become constrained during extremes of streamflow conditions or by operations dictated by environmental concerns. Many utilities have adopted integrated resource planning techniques to evaluate the economics of incremental generators added to a diverse portfolio of generators. Such techniques are invaluable in weighing the economics of incremental generating units with different fixed and variable costs, and that may have different effects on the operations and economics of the existing units.

Wind has its own peculiarities—generation that is variable, relatively uncontrollable, and less predictable than most other types of generation. Power system operations staff may be understandably vexed at the prospect of a resource over which they may have little or no control, and which they may feel cannot be predicted. Indeed, one real-time operator remarked to me that wind brings no value whatsoever to the power system. It is important to understand the true characteristics of wind (not entirely unpredictable, not entirely uncontrollable), and to acknowledge its limitations. While wind generation will never be the favored child of operators charged with constantly adjusting generation to meet load through time, it brings great value in reducing fossil-fuel consumption, associated atmospheric emissions, and dependence on imported fuels.

A complete understanding of the value of wind on an integrated power system is not a simple matter, and many utilities have undertaken complex studies, often with the help of specialized consulting firms, to help determine wind integration costs. The complexity of some of the studies, and the lack of uniformity of what is meant by integration costs, has contributed to some of the wide variation in results along with real differences in system characteristics. This book will provide some perspectives on the analysis, pointing out different methods that have been used, and pitfalls awaiting the unwary. It is important to realize that the subject is a very active area of investigation and improvement, and it is therefore likely that over time new and important approaches and results will be developed. The idea of this book is to help out anyone considering undertaking a wind valuation study for the first time, looking into improving a first- or second-generation analysis already performed, or monitoring the work of a consultant hired to perform a wind study. Areas that need additional attention are pointed out along the way.

Although the specifics of the interactions of a wind project and its interconnected power system can be complex, the analysis need not be overly complex. Every study of the behavior of power systems is an approximation. Determining a reasonable and sufficient level of modeling complexity is a key function of the competent analyst. Increased complexity does not necessarily result in increased accuracy for a variety of reasons. Probably the biggest reason is that the likelihood of error increases dramatically with complexity. Computer models of power system dispatch will contain many thousands of numbers representing the characteristics of the power system. Some percentage of these numbers will be wrong—possibly entered incorrectly, misinterpreted, or overlooked in the latest update, etc. Some errors may be tolerated, but some of them are likely capable of rendering the study completely invalid. In the press of time and staffing, the level of sophistication in error detection and study verification often lags to a dangerous degree.

In addition to outright data errors, model logic is often highly tuned to existing operations with traditional power generators, often making it necessary to change algorithm logic to correctly model a power system with substantial amounts of a resource with significantly different operating characteristics such as wind. In the complexity of modern power systems and computer models, changing the logic of one part of a computer model can have unexpected consequences, resulting in unrealistic simulations of actual operations in other areas.

In other words, given available staff and tools, modeling all the complexity inherent in a large power system's interaction with wind generation may not be the best approach for every study. There is no single best way to analyze wind on your power system. There are many approaches, and some are simpler than others. The simpler methods are not always inferior, and in some cases may well be superior to the more complex methods discussed. In most cases, simpler methods should be employed at least as a check on a more complex analysis. This book will try to be clear on the applicability of different methods.

One last note is that this subject is not only important from a social and environmental perspective, but it is also intellectually engaging and rewarding. Consider these intriguing challenges:

- The effect of a specific wind power project cannot be expressed separately from the power system into which it interconnects.
- The effects of one wind project can be markedly different from the effects of two or more acting together in a power system even if individually they act identically.
- The statistical behavior of wind generation cannot be easily or simply characterized.
- While average wind generation may be deduced from on-site wind speed measurements, the temporal behavior is a much more complicated matter.

There are interesting problems here, and the prospect of a non-polluting resource with no fuel costs is an enticement that makes the endeavor both socially and intellectually compelling.

REFERENCES

US Department of Energy (DOE) (2008). *20% Wind energy by 2030: Increasing wind energy's contribution to US electricity supply,* July. <http://www.20percentwind.org/20percent_wind_energy_report_revOct08.pdf>;.

World Wind Energy Association (WWEA) (2009). *World Wind Energy Report 2009.*

Chapter | two

Overview of System Impacts of Wind Generation on Power Systems

If a man does not know what port he is steering for, no wind is favorable to him.

Seneca, mid first century philosopher

Power systems typically consist of many interdependent components, including power-generating units, high-voltage transmission lines, substations, and customer loads. Adding new power plants to an existing portfolio of resources and loads can result in substantial operational changes for the pre-existing components. For example, adding a baseload thermal unit will reduce the frequency of dispatch for higher variable-cost resources such as natural gas powered generators. Similarly, adding a peaking unit may reduce the need to vary the output of intermediate units. It is not surprising then that the addition of wind generation to a power system similarly affects the operations of other power units in the power system. The effects of wind generation on a power system may be less familiar to the analyst and can be rather complex, potentially driving new operating procedures in order to maximize the economic efficiency of the joint power system that includes wind.

This chapter explores the effects of wind generation on other generators on an integrated power system in a qualitative sense, with the quantitative methods to follow in subsequent chapters. How wind affects other generation depends largely on the physical characteristics of the existing system, but also on the market structure of the power system—such elements as the length of the operating period (5, 10, 60 minutes, etc.), the

Valuing Wind Generation on Integrated Power Systems. DOI: 10.1016/B978-0-8155-2047-4.10002-X existence of liquid ancillary service markets, interconnection with other balancing area authorities, and interconnection with other liquid market points. For the purposes of this chapter, we will make the simplifying assumption of a vertically integrated utility with limited interconnection to other systems. Such systems do exist (e.g. Hawaii), and the more complex interactions involving other utilities or balancing areas through market mechanisms will be taken up in subsequent chapters.

2.1 PRIMARY ECONOMIC EFFECTS OF WIND POWER

Conceptually, treating wind generation as negative load is helpful in understanding the fundamental effect wind has on the balance of the power system generation. Because wind generation has no fuel costs and low variable operating costs, many analyses of the effects of wind on power systems treat wind as a must-run resource; contributing variability and uncertainty characteristics that are somewhat similar to load. Generation from other resources is adjusted in such a way as to meet the net load after subtracting wind generation. This is sometimes referred to as 'treating wind as negative load'. Figure 2.1 shows a typical week of wind and load for a power system, illustrating netting out the wind. The simplifying assumption that wind can be treated as a load deduction is reasonably valid for systems with relatively small amounts of interconnected wind generation.

The primary effect of wind on other power system generation stems from examining how other generators behave with and without the wind subtracted from the load. Economics dictates that there is a priority order of dispatching generating plants based on their variable costs. Generators with relatively low variable costs (fuel, and operation and maintenance) tend to be operated most of the time, and given the highest priority.

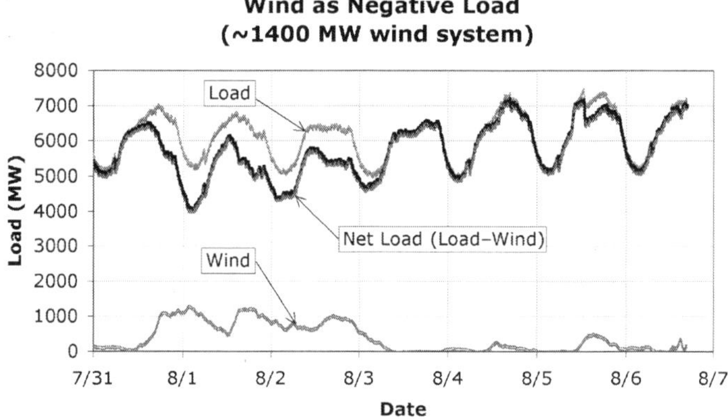

FIGURE 2.1 One week of load and wind on a system with about 1400 MW of wind.

Generators that experience significant savings when not operating will be given lower priority. The prioritized, or 'merit order', dispatch means power plants with the lowest variable costs are operated most frequently, and those with the highest variable costs are operated least frequently. In a generic sense, analysts group generators into three broad categories: peaking, intermediate load, and baseload units. Peaking units have the highest operating costs, baseload units the lowest operating costs, and intermediate load units are in between. These resources are 'stacked up' in their merit order as illustrated in Figure 2.2 to meet customer demand, with baseload generation at the bottom of the stack, intermediate next, and peaking generation on top as needed. The ordering of generating unit dispatch minimizes overall operating costs.

Much of the value of wind power is derived from the savings in operating costs and emissions associated with reducing generation from peaking and intermediate load units, as illustrated in Figure 2.2.[1] Peaking units are usually made up of lower efficiency single-cycle combustion turbines, or highly inefficient fossil-fueled steam boilers. The most expensive power plants to operate are energized with lowest priority in order to keep costs down. To the extent that wind generation causes greater sales or reduced purchases from liquid markets, the savings may alternatively be tied to prevailing market prices at the time the wind generation occurs (saving variable costs on generation from other power systems). The ability to take advantage of markets depends strongly on the market structure and the ability to forecast wind generation ahead of time. Utilities find they may be able to sell some portion of the wind generation into the sub-hourly market or forward (hour-, day-, or week-ahead) markets depending on the availability and accuracy of wind forecasts.

2.2 ROLE OF WIND FORECASTS IN WIND POWER ECONOMICS

The importance of wind forecasts is widely touted, but often poorly understood. Schedules of estimated loads and power generation are

[1] It should be noted that the situation for power systems with hydro resources could be somewhat different. Hydro may be broken into baseload, intermediate load, and peaking components. Savings in energy dispatched from hydro units may accrue as water in reservoirs that may be released at a later time to displace thermal generation at other times, or for market sales to other systems that accomplish the same. In addition, coal plants are generally considered low-variable cost, baseload generation that are not primarily displaced by wind (one study in Texas estimated that each megawatt-hour of wind energy displaced 0.19 megawatt-hours of coal (Cullen, 2008)). This situation could change dramatically if a carbon dioxide penalty (e.g. a tax) is levied on coal plants. In that case, the position of coal in the merit order dispatch could change, resulting in wind energy displacing more coal generation than is currently the case in the USA. Displacement of coal may also become more prevalent at higher wind penetration levels considered in Chapter 14.

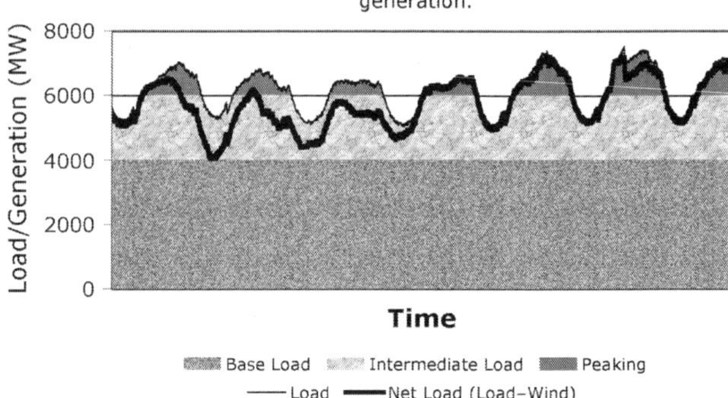

FIGURE 2.2 Resources needed to meet the load without wind (top line) and after taking the wind into account. The difference between the lines represents power generation from other sources that is not needed due to the presence of wind on the system.

developed a day in advance, and improved estimates are provided typically an hour in advance of the operating period, and finalized 15 minutes to 2 hours before the beginning of the operating period. Accurate schedules reduce the need for power systems to maintain the ability to increase or decrease generation due to unaccounted changes in wind output or system demand. Responding to unforecast load or generation is accomplished primarily by especially flexible reserve generating capability. Conversely, variability of wind and load that is reflected in the schedules can be accommodated using the most cost-effective means available. The accuracy of wind forecasts that inform wind schedules is an important factor in determining the level of reserve generation needed, and hence importantly contribute to the value of the wind generation.

Most utilities have experience with forecasting loads hours, days, weeks, and even years ahead. Forecasts for up to a week or two are usually based on weather models and historical behavior of loads. Electric demand is a strong function of temperatures, wind speed, and humidity. Other factors are also important, such as hour of the day, day of the week and year (e.g. weekends or holidays), and industrial load characteristics (e.g. dependence of irrigation loads on time of year and precipitation). Weather models take into account prevailing temperatures, pressures, wind speeds, insolation, ground reflectivity, and other relevant data to predict the movement of air using physics-based equations of motion. These same models are also used to predict wind speed and direction. The data can be translated into power generation using information specific to a wind project, such as power curves and unit availability.

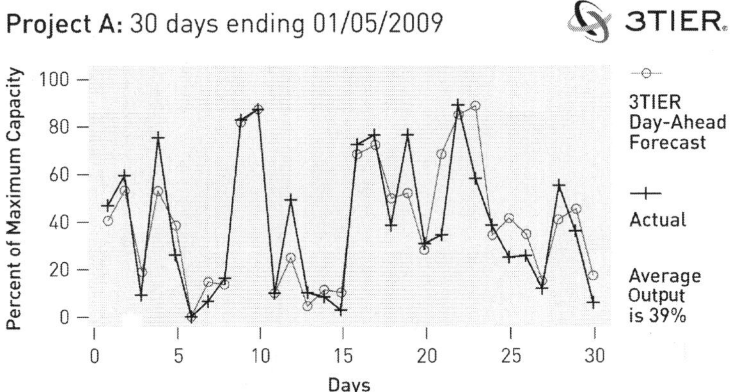

FIGURE 2.3 Commercially available wind forecast made one day ahead of actual operation for a 31-day period. *Courtesy of 3Tier Environmental Forecast Group.*

There are various levels of sophistication of wind power forecasts, and different approaches for different timeframes that will be described in more detail in a later chapter. Suffice it to say that the value of wind energy is at least somewhat dependent on the accuracy with which the wind can be forecast. Figure 2.3 shows an example of the accuracy of a commercially available day-ahead wind forecast over a month for one wind farm. The accuracy of wind forecasts may vary by wind-turbine site, skill of the forecast provider (which usually improves over time for new sites), time of year, and timeframe (hour ahead, day ahead, etc.).

2.3 WIND AS AN ENERGY RESOURCE

Power planners usually calculate the need for new resources based on the power system's capability of meeting the peak system demand[2] during weather extremes when demand is especially high. Wind power provides a relatively small benefit in meeting peak demand compared to fuel cost savings. Some utilities assume wind can meet none of the power system demand; others may use a value closer to the observed wind generation capacity factor during peak demand periods. The most accurate method for determining any generator's contribution to meeting peak loads is to determine its peak load-carrying capability. Chapter 10 addresses wind's contribution to meeting peak loads in detail.

While the primary value proposition for wind is for fuel displacement, or a combination of fuel displacement and market effects, there are countervailing costs to consider as well. There is of course the cost of the

[2] Power systems dominated by hydro power often plan on an energy basis, looking at extended drought conditions under which the energy from the hydro system is the limiting factor as opposed to the peak hour capability.

wind generation itself, but we will take that as a given[3] for now. Power systems must carry sufficient flexibility to increase or decrease generation as needed for unexpected changes in demand or resource performance such as the failure of a large generator. Flexibility to increase or decrease generating capability on short notice is often referred to as 'reserve capability' or 'reserve margin', or simply 'reserves'. There are different kinds of reserves held for different purposes and timeframes. The breakdown and description of various reserve categories is described in greater detail in later chapters. For now, it is most important to note that because wind adds both variability and uncertainty to the net power balance of a power system, there is an increased need to hold reserve capability for systems adding wind power. Calculating the amount and cost of providing the additional reserves needed to accommodate wind added to a power system is the primary focus of this book.

Most power systems have a minimum operating period over which load and generation schedules are submitted. Operating periods range from as short as 5 minutes to as long as an hour. Power transactions from generator to load are fixed over the operating period for most of the generation. The fixed amounts of energy expected to be generated and delivered to load over the operating period are submitted to transmission providers as 'schedules'. Within the operating period the actual performance of generation and demand deviates from the schedule. Maintaining system reliability standards requires that the differences between schedules and actual performance must be met by increasing or decreasing generation, as needed, on reserve generating units. Figure 2.4 illustrates schedules versus actual operations.[4]

There are costs associated with both maintaining generating units on reserve status, and with operating those units—either backing them down when the load net of wind is lower than the scheduled generation, or increasing generation from reserve units if the scheduled generation falls short. Both effects can occur within any given operating period. Costs of backing units down can accrue from reduced generation efficiency, and increased maintenance due to wear and tear on machinery undergoing more frequent and more dynamic changes in torque on turbine and generator components. Hydro systems can also experience increased costs due to de-optimized use of water. Increasing generation from reserve units results in additional operating (fuel)

---

[3] This book is primarily about the value of wind power in the context of modern power systems. The detailed cost of constructing and financing wind projects is beyond the present scope, involving not only the prevailing cost of wind power plant components (wind turbines, transformers, etc.) and quality of the wind resource, but also tax treatment, cost of money, and any tax incentives that may be available.

[4] For simplicity, the variability discussed here is confined to load and wind generation. In practice, virtually all generation varies appreciably through time. That effect is generally thought to be small compared to wind and load variability.

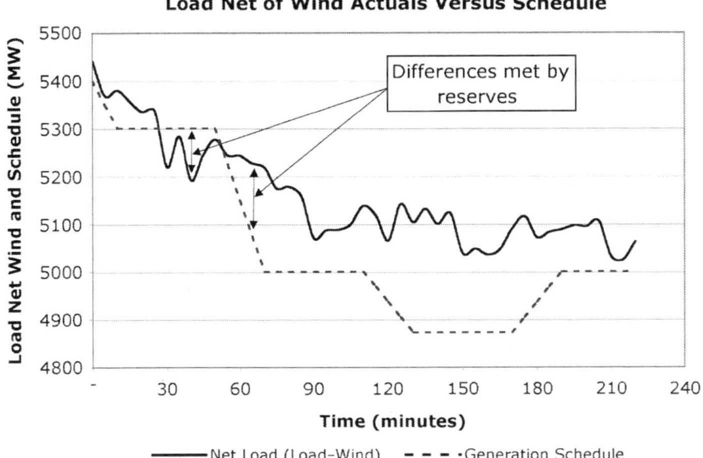

FIGURE 2.4 Difference between the scheduled generation and actual load (net of wind).

costs, as well as potentially lower efficiency and higher maintenance costs.

On at least a conceptual basis, the reserves needed to accommodate wind on the power system could exceed the contribution to meeting peak loads that wind brings. This suggests that wind could, in some special circumstances (Kirby & Milligan, 2008), increase the need for generating capability to meet peak loads. In North America and northern Europe, this probably never occurs because periods of peak demand normally occur at times of weather extremes, and weather extremes are usually accompanied by large-scale high-pressure systems that tend to idle wind generation. With wind generation idled, there is little need to hold reserve generation. However, many utilities may not be prepared to hold varying amounts of reserves (e.g. by time of day, or by wind output level). For those utilities, changes in operating procedures are likely needed to efficiently accommodate significant amounts of wind. This effect is described in greater detail later.

2.4 OTHER POTENTIALLY IMPORTANT EFFECTS

A number of wind impacts on power systems are not generally important cost-causation mechanisms when wind represents a small percentage of power system generation, but need to be noted by power system planners and analysts on systems with significant amounts of wind. Figure 2.5 illustrates some of the challenges associated with a power system with 20% of the energy generation coming from wind. Power systems with high levels of wind generally experience periods in which the overall wind generation is high enough to cause the net load to be quite small—

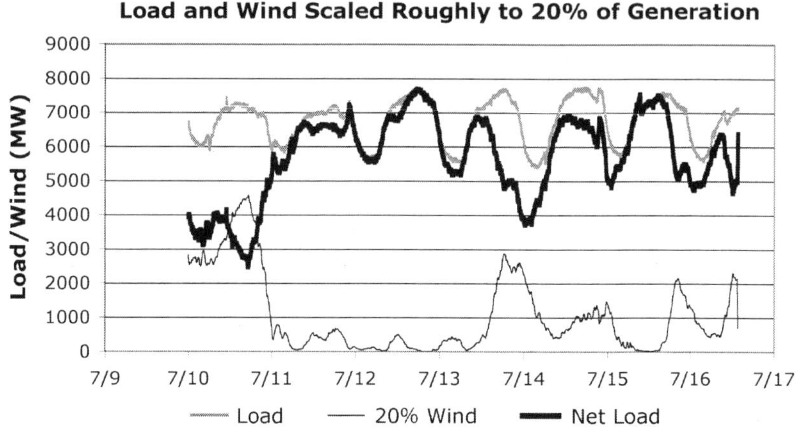

FIGURE 2.5 Example of system net load with 20% energy production met by wind.

potentially smaller than the utility's minimum generating requirements. If the power cannot be marketed to neighboring systems, or absorbed in some other way, the wind turbines may have to be limited and the generation lost. Wind-turbine technology allows for a range of flexible operations such as operating at a fixed percentage of output capability (at a given wind speed), or operated at a fixed amount of power (MW) below output capability. The rate of increase in output can be limited, and the total output can also be limited.

These capabilities suggest that treating wind simply as negative load is an oversimplification—wind generation can in fact be actively managed as necessary. For systems with large amounts of wind in which these capabilities (i.e. limiting wind output) are important and used on a routine basis, it is important to treat wind more as a partially dispatchable resource.

A period of significant wind generation that may have to be limited is illustrated in Figure 2.5, between the date markers 7/10 and 7/11, and again near the 7/14 date marker. These are periods in which the wind generation on a system-wide basis is quite high, driving net load to low levels that it may not be practical to accommodate. For example, some thermal units may not be able to operate reliably below approximately 50% of nameplate generation. Taking a unit down completely may result in its being unavailable for the next morning's load ramp when the wind may have died down. To avoid such a situation, the generation from the wind may be limited for short periods of time. Systems with large amounts of wind may investigate engineering changes to thermal generating units to improve the ability to change their output and reduce below existing minimum generating limits.

Another effect of potential significance is illustrated in the 7/10–11 event depicted in Figure 2.5, where a steep and prolonged period of generation escalation from other resources is required to balance the rapid and sustained reduction in wind generation. Planners need to estimate the ramp rate requirements that significant amounts of wind can bring. Wind can fall off relatively rapidly, and though sufficient generating capability may be available from otherwise idled units, those units may not be able to increase generation fast enough to compensate for the falling wind. In such cases, the wind may need to be limited, or dispatchable generators added or modified, to provide sufficient ramping capability. A similar issue may occur in dropping generation as wind rises rapidly, although this seems to be a much less important effect on thermal-based power systems.[5] It should be kept in mind that extreme ramping events over a reasonably geographically diverse set of wind projects are relatively infrequent, and the effects on the power system ameliorated by more refined forecasting techniques.

Power system reliability issues can also become important on systems with significant amounts of wind. Early wind turbines were designed to shut down with a loss of station service. This can have the effect of exacerbating what might otherwise be momentary transmission outages (e.g. due to trees blown into transmission lines or lightning strikes). Reliability organizations are beginning to require wind generation to adhere to new low-voltage ride-through requirements that maintain wind-turbine operations through momentary losses of line voltage. Some wind turbines are capable of providing dynamic, or semi-dynamic, reactive support to the transmission systems as well. It may also be necessary to reinforce the existing transmission and distribution systems to ensure that wind energy does not cause localized reliability issues. While important to reliability considerations, transmission and distribution system reinforcement, low-voltage ride-through, and reactive support are not considered in detail here.

2.5 PROPERTIES OF WIND OUTPUT IN AGGREGATE

It is important to understand some general characteristics of wind generation in aggregate over a power system. While the power consumption in a single household is extremely variable and difficult to predict, the consumption of a large number of households summed together can be forecast with considerable accuracy. The same is true with respect to forecasting individual wind turbines versus wind turbines collected together in a wind project—the output of a wind project (many wind turbines) can be much more accurately forecast than a single turbine, or

[5] The rapid rise in wind generation as weather fronts pass through, necessitating reduced generation from other generating units, can be the primary wind integration cost driver in predominantly hydro systems.

similarly, the output of the collected wind projects on a power system. Mathematically, much of the variability and randomness cancels out—for example, when one person turns on a light switch for some unknown and unpredictable reason, another person may be simultaneously turning out a light—the individual actions are unimportant, and only the net effect matters to power system operators.

Similarly, in large power systems the variability and predictability of individual wind turbines is unimportant from a total system perspective. It is only the aggregate behavior that matters at the integrated power system level.[6] Just as with household demand, the aggregate behavior of many wind projects is less variable and more predictable than individual wind projects or a single wind turbine. A more detailed examination of the behavior of wind project output will be undertaken in later chapters, but it is important to understand some of the general characteristics of wind power at the outset.

Wind is the result of differential heating of the earth's surface, and the root of the energy resource itself is actually the sun. The differential heating can occur on a large-scale basis such as the difference in solar radiation impinging on the poles versus near the equator, or more localized phenomena such as occur between bodies of land and water. The very uneven heating of the earth's surface due to differences in earth reflectivity (forest versus desert, plowed farmland versus fallow, etc.) contributes to producing wind.

The specific behavior of wind is extraordinarily complex. Forecasts of wind speed, and hence of wind power output at wind projects, are usually based on large-scale weather models—so-called meso-scale weather models. These models take into account real-time measurements of physical properties of the atmosphere, earth surface, and even ocean conditions. Data are input into complex computer models that use the mathematics and physics of hydrodynamics and solar influx to project the behavior of the atmosphere forward in time. It is rather famously known that chaotic processes limit the ultimate accuracy of meso-scale models (the so-called 'butterfly effect'). Nevertheless, forecast technology has not reached its theoretical limit, and the accuracy of the models continues to improve with faster computer technology and more detailed real-time data feeds from ground stations and satellites.

Weather models generally do a better job of forecasting larger-scale phenomena, such as high and low daily temperatures, than they do on more localized phenomena such as precipitation. Wind falls between

---

[6] There is an inherent assumption that sufficient power transmission exists to allow the wind projects to present their aggregated output to the power system. Wind projects on weak transmission grids or interconnections can present additional localized effects that may be specific to one or a few islanded wind projects. This book assumes a well-integrated power system with no significant transmission congestion.

these two extremes, with weather models doing a generally better job forecasting the wind speed and direction at a wind site than precipitation, but not as accurately as daily temperatures. In addition, the accuracy of wind speed forecasts tends to be better for the next day than they are for several days out.

A time scale of great importance to power systems is the next few minutes to hours. Meso-scale weather models are of limited help on that time scale without specialized modification, and potentially hybridization with other data and modeling techniques. The reason for this is that meso-scale models take massive amounts of data from weather stations that are generally available no more frequently than hourly, and the models themselves may take several hours to process the data.

Despite wind's reputation for variability, the output of a wind project may not vary more than a few percent over a period of an hour, and often the best forecast of what the wind generation will be in the next 2 hours is simply the current level of output. This type of forecast—assuming the current level of generation will continue—is called a 'persistence forecast'. Improving on persistence forecasts is possible and can involve a variety of techniques ranging from employing relatively simple time series analysis, to involving real-time data from wind projects and meteorological stations combined in complex ways with information from meso-scale models.

Of great concern is the extent to which wind generation from separate projects may vary in concert with one another. If all the wind projects on an integrated power system increase and decrease generation levels simultaneously, the effect on the power system is very much greater than if there is a more random relationship among the wind projects. The effect can be quantified with the statistical correlation function. The correlation function assigns 1.0 to pairs of data sets that move perfectly in unison, and −1.0 to data sets that move completely opposite (when one increases generation, the other decreases and vice versa). Correlations between wind projects are generally higher on longer time scales than shorter ones, and correlations between wind projects that are geographically near one another are generally higher (closer to 1.0) than those that are more distant. These relationships are illustrated in Figure 2.6.

On the shortest time scales of a few minutes or less, wind project output is generally quite independent from one another. Power system operators hold regulating reserves to accommodate the fluctuations of demand over periods of a few seconds to a few minutes. Because of the relatively uncorrelated nature of fluctuations from wind projects on that time scale, the increased regulating reserve requirements for wind projects is relatively small compared with the longer time scale reserve requirements. The incremental regulating reserve requirement for wind is typically of the order of 1% of the total reserve requirement.

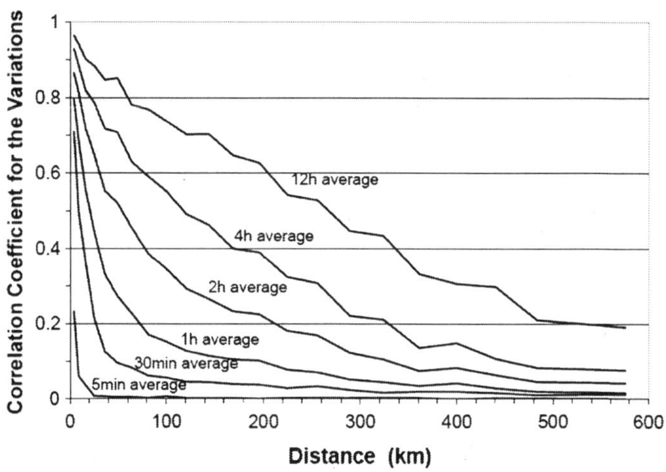

FIGURE 2.6 Correlations among changes in wind project output on differing times scales and at a range of distances from one another. *Source: Ernst (1999). This figure was created and prepared by the National Renewable Energy Laboratory for the US Department of Energy.*

As Figure 2.6 suggests, there are statistically significant relationships between even distantly located wind projects on time scales of an hour or more. From a power systems operations point of view, this is somewhat unfortunate—if the output of projects were completely uncorrelated, the output of a sufficiently large fleet of wind projects would approach a relatively constant level that would be relatively easily predicted and managed. Although local topographical features often dominate wind project output, larger-scale weather drivers are also important. There are times when large-scale weather systems can dominate the weather pattern over a large fraction of a continent. Of particular concern are large high-pressure systems and fast-moving weather fronts.

2.5.1 Effects of high-pressure systems and weather fronts

Wind may have daily or seasonal patterns that match up either well, or not so well, with demand for electric power. There are wind regimes that produce most of their generation from movements of winter storm systems, others that are driven more by the differential heating between land and sea in the summer time, and still others that benefit more substantially from the unsettled weather found in spring and fall. Some wind regimes show strong diurnal patterns, some do not. Some of the diurnal patterns result in greater daytime generation, while others result in greater generation at night. There is no single characterization of the wind that fits all sites. Nevertheless, there are important relationships between wind and high-load events. Extremes in demand occur during temperature extremes (hot, or cold, or both), and those extremes are usually accompanied by large-scale high-pressure systems that have a fairly predictable effect on wind.

High-pressure systems are air masses in which the air pressure is higher than the surrounding areas. They are associated with light winds and little cloud cover. Temperature extremes (hot summer days, cold winter days) that drive demand for electric power often accompany high-pressure systems. In the summer, the low cloud cover and light winds contribute to high-temperature extremes. In the winter, high-pressure systems contribute to low-temperature extremes from the increased radiative cooling at night, also due to low cloud cover. This is an important effect because it suggests that periods when electric demand is high will often be accompanied by little or no power generated from wind projects—and the effect can hold for wind projects spaced even hundreds of kilometers from one another within the largest high-pressure systems. Such a system is illustrated in Figure 2.7.

The high-pressure system of 12–14 January 2007 illustrated in Figure 2.7 was accompanied by a high-load event in the winter-peaking Pacific Northwest. Figure 2.8 illustrates the aggregate wind production on the Bonneville Power Administration power system during that week.

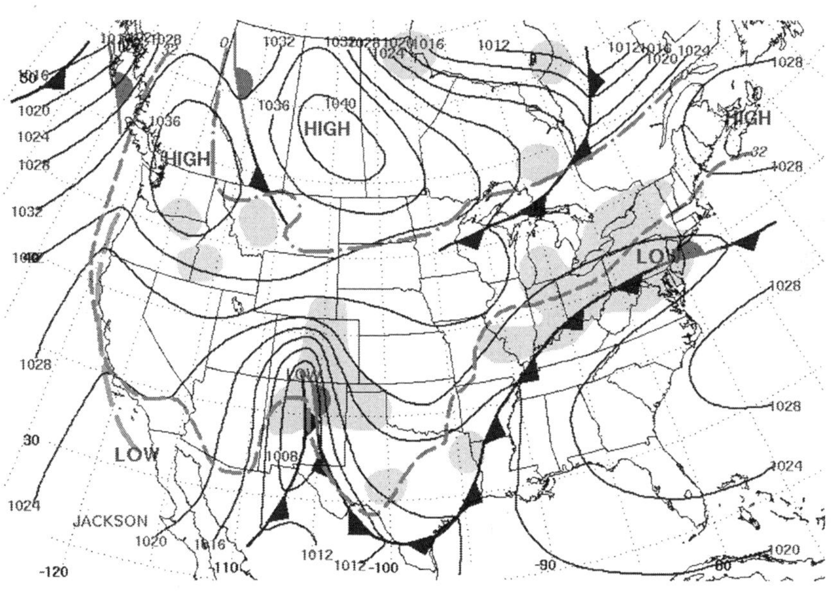

FIGURE 2.7 Surface pressure map illustrating a large-scale high-pressure system over the northwestern part of the USA and in central Canada for 14 January 2007. The system was accompanied by calm winds and high electrical demand in the Pacific Northwest. Shaded areas indicate precipitation over the past 24 hours. *Source: National Oceanographic and Atmospheric Administration, <http://www.hpc.ncep.noaa.gov/dailywxmap/index_20070114.html>.*

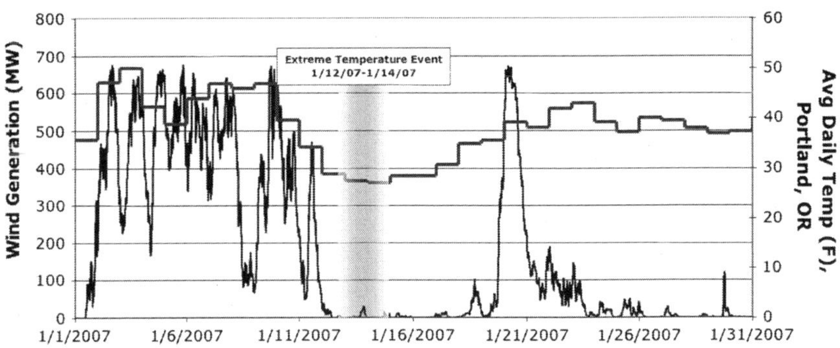

FIGURE 2.8 Wind generation on the Bonneville Power Administration system in the Pacific Northwest (USA) throughout the 12–14 January 2007 high-pressure event depicted in Figure 2.7. Winds tend to be calm during high-pressure events, while demand for power is likely to be highest. Data represent eight wind projects totaling 722 MW spanning approximately 200 km.

Winds at the periphery of high-pressure systems can be relatively high, so at the onset and ending of high-pressure events there may be considerable wind production. In the midst of the high-pressure event itself, when power demand may be at its height, winds are likely to be relatively calm.

2.5.2 Weather fronts and wind ramps

Rapid changes in wind output can be caused by the passage of weather fronts. Weather fronts are the boundaries between air masses with different densities (different temperatures or humidity). Passing weather fronts are usually accompanied by changes in wind velocity. In the northern hemisphere, weather fronts often move from west to east and can cause rapid changes in wind output from even geographically distant wind projects, especially wind projects oriented along a roughly north–south line. The severity and frequency of wind ramps tend to be reduced for systems with wind projects dispersed over large regions.

Wind-ramping events can cause the most severe impacts on other generators. Most power system operators are well aware of severe weather conditions that drive peak-load events. Operators may be less accustomed to remaining vigilant to the sudden onset of wind-ramp events driven by passing weather fronts, especially those not likely to result in peak-load events. These types of events will increase in significance to operators of systems with large amounts of wind generation, and also to forecasters historically more attuned to accurately predicting high and low temperatures. Weather models are not typically 'tuned' to forecasting these types of events with great accuracy, and specialized versions are expected to appear in order to more accurately forecast wind energy. Because they tend to move in predictable directions, it may be both possible and desirable to establish strategically placed meteorological stations to collect data upwind of (west in the northern hemisphere) wind projects to better detect and forecast weather fronts.

2.5.3 Wind generation data

One of the greatest challenges for analysts preparing to examine the economic effects of prospective wind generation is developing credible representations of wind output for wind projects that have yet to be constructed. Analysts may not have a clear view of exactly where the new wind projects are likely to be sited, making the problem especially difficult. Methods available to analysts for preparing reasonable approximations of wind behavior are detailed in Chapter 4.

One approach to estimating the effects of additional wind projects is to start with wind generation data that may already be available from existing projects on the power system. While this is a reasonable starting point, care must be taken to preserve the complex relationships among multiple wind power projects acting in aggregate. For example, the behavior of two 100-MW wind projects cannot normally be adequately represented by simply scaling up the output of a 100-MW project and multiplying by 2. Again, the reason this is inaccurate is that two wind projects, even two physically adjacent wind projects, will not have identical generation through time—especially on the shortest timescales. Some analysts have attempted to use time shifts with scaling to represent new wind projects at some distance (proportional to the time shift) from existing projects. This can be a useful tool, but again care must be taken to ensure that the pertinent relationships, as illustrated in Figure 2.6, are maintained. More on this important subject in Chapter 4.

2.6 SUMMARY

Understanding the value of wind entails addressing both the primary value proposition (reduced fuel costs and/or changes in market transactions) as well as the offsetting costs associated with the increased need to hold and dispatch reserve units. The offsetting costs are often referred to as 'wind integration costs'. Wind integration costs may be used by transmission providers to establish wind integration tariffs or, more typically, by utilities purchasing wind generation and needing to understand the value of wind power in comparison to competing technologies.

Calculating the primary value proposition of wind energy is relatively straightforward, requiring information about the timing of the wind generation and either market price forecasts or computer dispatch model commonly employed by utilities for power planning and dispatch decisions. The extent to which wind may contribute to meeting peak demand on power systems can be determined using techniques that have been applied to other power generation technologies. Calculating wind integration costs is a relatively more complex problem, depending on not just the amount and average timing of the wind generation, but also its variability and predictability. The behavior of wind generation is not easily modeled and generally requires data from projects that are not yet built, making data acquisition and analysis difficult. In addition, the effects of wind generation are typically not linear—in other words, doubling the wind generating capability on a system does not necessarily double the need or cost of holding and dispatching reserve units. While the potential difficulties are great, utilities are finding a variety of ways to effectively address them.

REFERENCES

Cullen, J. (2008). *What's powering wind? Measuring the environmental benefits of wind generated electricity.* 2008 Annual Meeting, American Agricultural Economics Association (new name 2008: Agricultural and Applied Economics Association).

Ernst, B. (1999). *Analysis of wind power ancillary services characteristics with German 250 MW wind data.* NREL Report No. TP-500-26969. 38 p. Available at: <http://www.nrel.gov/publications/>

Kirby, B., & Milligan, M. (2008). An examination of capacity and ramping impacts of wind energy on power systems. *Electricity Journal,* Issue 7.

Chapter three

General Approaches to Valuing Wind on Power Systems

The winds and waves are always on the side of the ablest navigators.

Edward Gibbon, 1737–1794, English historian and Member of Parliament

Before the advent of integrated resource planning, electric utility analysts evaluated the net benefit or cost of a resource in isolation from the rest of the power system. Planners would start with an assumed operation of the power plant—whether it would operate all the time as a baseload unit, or relatively infrequently for peaking, or for intermediate load service. Given the operating profile, the analyst would tally the fuel and operation and maintenance (O&M) costs, which would then be added to the cost of acquiring and constructing the power plant. The cost of financing, as well as tax and accounting ramifications, would also be taken into account, to determine a total cost for the generated power, often expressed in terms of dollars per megawatt-hour. Those costs would be compared across other alternatives—either other power plant options, or possibly contractual purchases of power. The least-cost alternative would normally be selected for construction, or at least further consideration given to competing environmental and regulatory concerns.

Vertically integrated utilities primarily seek the least-cost resource that provides sufficient generating capability (energy and peaking capability) to serve customer demand. The economic evaluation for an independent power producer may be done on a basis different from described above. For an independent power producer, the decision to purchase or construct a power plant is not based on adequacy of supply to meet customer demand, but is a business decision that is more usually based on comparing the return on investment for competing business

Valuing Wind Generation on Integrated Power Systems. DOI: 10.1016/B978-0-8155-2047-4.10003-1

opportunities. The cost side of the analysis is generally similar to that described above for an electric utility, but more attention is given to the value of the power produced. The value of the power may be represented by the purchase price in a proposed multi-year power sales contract, or the value may be derived from forecasted wholesale market prices for electric power. Forecasts of forward market prices entail much uncertainty, and the analysis is also complicated by the need to estimate fuel costs that may also be quite uncertain and volatile (especially natural gas). The determination to go forward with a purchase or construction opportunity entails a detailed analysis of the risks and benefits, and ultimately whether the opportunity provides a return on investment that is attractive to the particular entity doing the calculation.

In the past three decades or so, utilities moved away from evaluating the costs of power plant opportunities in isolation, to a more system-wide analysis of the value of prospective generation—the so-called integrated resource planning approach. It became increasingly apparent that the addition of any new generating unit affected the operations and cost-effectiveness of other generating units on the power system. Generally, the addition of one power unit would reduce the dispatch of other units and the relevant measure of cost-effectiveness should be the overall cost of the portfolio of generating resources. In other words, the cost-effectiveness test shifted away from the point of view of the individual last-added power plant, to the cost of the overall system to the ultimate consumers. Shifting to a system-wide evaluation necessitated implementing complex power system models capable of assessing the interactions among power plants within a utility and also the interaction with the power plants of other utilities through purchases and sales in the wholesale marketplace.

Integrated resource planning (IRP) studies involve modeling the behavior of all the generators in the utility's portfolio (utility-owned power plants and contracted purchases and sales) in the context of the utility's native demand as well as a representation of access to market purchases and sales in monthly, daily, and sub-daily markets. Generally, this involves a multi-year analysis in which various alternatives for meeting customer demand are tested. These studies usually involve computer programs that seek to optimize the operations of available generators and markets to meet demand by minimizing costs and maximizing wholesale revenues. Most of the models move sequentially through time—usually an hour at a time. In each time period, the model establishes generation levels for each power plant sufficient to meet load and other criteria (emissions constraints, reserve margins, etc.) while minimizing costs. Operating costs are tallied for each operating period, along with purchase power costs and wholesale market sales revenues.

The end result is an annual stream of costs that are usually present-valued at the utility's weighted average cost of capital to establish the revenue requirement over the years of the study horizon. Multiple studies are run with alternative portfolios of resources (one or more added power plants) and the present value revenue requirement (PVRR) can be compared across resource alternatives.

This short description is meant to set the context for valuing wind on integrated power systems, and is not a detailed description of IRP analysis. IRP analyses are very complex, involving large amounts of data about existing and potential power plant additions, the growth of customer demand, end-effects when the study horizon does not match the economic life of the alternative resources, etc. A full accounting of IRP methods (Bolinger & Wiser, 2005) is beyond the scope of this book, but it is important to understand the basic ideas as the accurate evaluation of wind power costs necessarily entails examining the effect of wind on other resources in the power system.

Valuing wind on power systems can be done using IRP methods—running computer dispatch models with incremental wind generation on the system, and alternatively without the wind generation. The difference in PVRR is the net cost or benefit associated with wind. Although this sounds enticingly simple, especially for utilities that already use dispatch models and perform IRP studies, there are some important complicating factors that need to be adequately addressed, and which are described in detail in this and the following chapters. As wind generation becomes a significant supplier of power system energy, there is an increased need to hold and dispatch reserves from more flexible generators on the power system. There may also be more energy trading with neighboring systems and attendant market transaction costs. It is important to capture those costs in some way. Including wind in complex dispatch models for IRP analysis is one way, but not the only way to do this valuation.

3.1 WIND VALUATION COMPONENTS

There are six general categories of costs and values that form the core of a thorough wind valuation analysis:

1. Direct cost of the wind power.
2. Gross value of the generated energy.
3. Value of renewable energy credits and emissions reductions.
4. Cost of holding additional reserves due to wind variability and uncertainty.
5. Effects on reserve generation operating costs.
6. Other effects on balance of system generators and increased trading costs.

The somewhat novel challenge associated specifically with wind energy is in determining costs associated with reserve requirements—items 4, 5, and 6 above. This chapter is a brief review of methods associated with each of the relatively straightforward valuation components. Perhaps the biggest challenge associated with traditional resource valuation is gathering current market data. The current or future cost of wind turbines, wholesale electric market prices, and renewable energy credit value is a very dynamic world. Nevertheless, those issues are similar in many respects to the uncertainties facing other electric generation technologies. Defining, quantifying, and establishing the cost of reserves provided for wind generation is a relatively new challenge for most analysts and represents the bulk of the technical challenge in valuing wind power.

3.1.1 Direct wind generation cost

The cost of wind power generation may be as simple as the price stipulated in a long-term power purchase agreement that a utility may have entered into, or may be offered as part of a bidding process. For utilities undertaking to construct and operate their own wind power project, the analysis is more complex. Some of the costs that make up the total cost of a wind power plant are similar to those of other power-generating stations: site preparation and licensing; leasing agreements and royalty payments to land owners; capital outlays for turbines, towers, transformers, and interconnection equipment (which may include a substation and high-voltage transmission lines); wheeling and ancillary service costs to move the power from where it is generated to the receiving entity; and ongoing O&M costs.

The cost of wind power rose steadily from 2000 through 2008, despite steadily improving technology. Commodity prices for everything from concrete, copper, and steel, to basic transportation costs accounted for much of the increases. There have also been shortages of major components such as the turbines themselves that may have contributed to the price increases. Costs for competing technologies have also increased with commodity prices, and it is unclear whether the cost of wind generation has increased faster than competing thermal generation. Figure 3.1 shows historical wind project installed costs (capital and construction costs, excluding O&M) in the USA. Nominal wind installation costs in 2008, as of the time of this writing, appear to be in the range of $2300–2500 per kilowatt (nominal) in the USA. With rapidly changing economic conditions, it is difficult to make prognostications about the next few years, though the economic downturn of late 2008 appears to be ameliorating the upward cost pressures experienced from 2000 to 2008.

O&M cost trends are shown in Figure 3.2. Absent fuel costs, the predominant costs of wind projects derive from the capital and construction costs. To develop costs on an energy basis, the O&M costs are added to annualized installation costs spread over the expected annual energy generation. As an example, assume a wind project with an installed cost of $2500/kW represents $300,000 per year repayment (20 years at

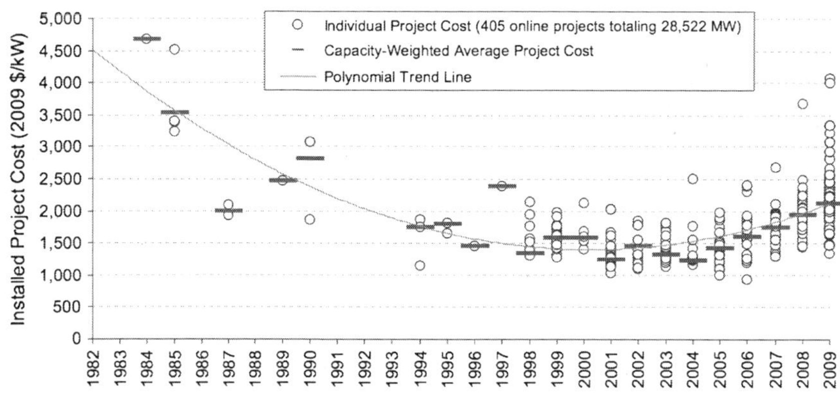

FIGURE 3.1 Historical wind project installation costs in the USA. *Source: 2009 Wind Technologies Market Report, US Department of Energy Office of Energy Efficiency and Renewable Energy.*

10.5% discount rate) and producing 2600 MWh per year for each MW of installed generation (30% capacity factor) has O&M costs of $15/MWh. Spreading $300,000 over 2600 MWh yields $115/MWh in nominal dollar terms. Adding O&M costs results in an all-in cost of $130/MWh.

The example above illustrates the dependence of the energy cost basis on the amount of energy produced. If the same project cost assumptions are applied to a project producing 3000 MWh per year (35% capacity factor), the cost falls from $130/MWh calculated above to $115/MWh. Commercial-scale wind project capacity factors vary from the mid teens to more than 60%. Offshore projects may be even higher. Onshore

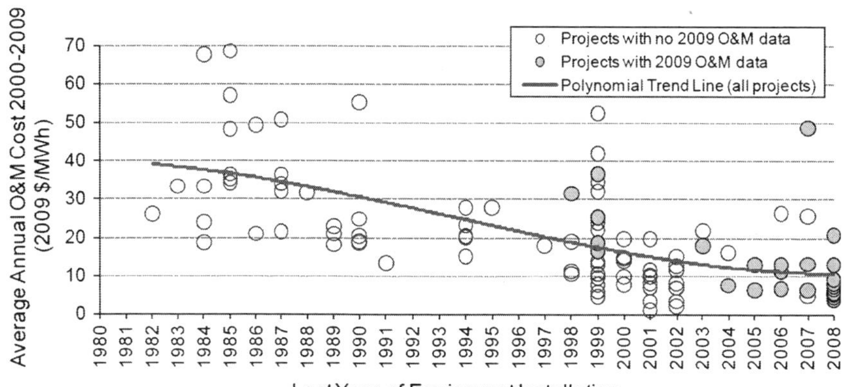

FIGURE 3.2 Historical trends in O&M costs in the USA. *Source: 2009 Wind Technologies Market Report, US Department of Energy Office of Energy Efficiency and Renewable Energy.*

projects in the USA tend to range from the mid twenties to the high thirties.

There may be offsetting cost considerations, such as tax credits, feed-in tariffs, and the value of renewable energy credits (RECs) produced. Consideration of REC value is considered as a separate value category below. Federal legislation in the USA has provided a production tax credit (PTC) of roughly $20/MWh for wind energy production over the first 10 years of project life. The legislation comes up for renewal every year or two, and it is difficult to predict what the long-term prospects are for the continuation of the tax credits. Some European countries have opted more for feed-in tariffs—guaranteed long-term payments for power produced from certain qualifying facilities such as wind and solar to ensure economic viability of those technologies.

3.1.2 Gross value of generated energy

Wind generation can reasonably be valued at prevailing market prices for utilities with access to liquid market points. Historical prices may be obtained from many of the Independent System Operators (e.g. AESO, CAISO, ERCOT, MISO, PJM, NYISO, Nord Pool, etc.). Commercially available wholesale price forecasts may be purchased from third parties, or may be constructed by the utilities themselves. These are generally based on estimated fuel costs in the relevant market regions, forecasts of load growth in those regions, and representations of existing and projected resource additions. Purchased forecasts, or those produced internally, are needed for resource planning purposes to the extent a utility may be engaged in planning processes. Hourly or daily market prices can be used to produce a value for the wind energy produced. Alternatively, wind generation may be incorporated directly into planning models to determine the value in the same manner as may be done for other resources. Such studies will explicitly value the reduction in fuel costs of other resources in the power system, and potentially reduction in emission-related costs as well. Chapter 8 contains a much more detailed description of this process.

Calculating the value of wind generation starts with developing a profile of expected wind generation by hour of the day and month (or at least season) of the year. Calculating the value of the wind generation is then a relatively simple matter of multiplying the expected cost on any hour by the expected price for that hour. This method breaks down if the amount of wind is largely compared to the existing markets or generating capability of the system. For example, if the wind in a region were to increase to a level where it becomes a significant amount of the energy traded on an hour, it would be expected to depress the market-clearing price. In such cases it will be necessary to employ more sophisticated dispatch and market models to estimate the value of the wind generation.

Previously noted was the difficulty in producing wind data for projects that may not yet be built for the purpose of determining reserve requirements. Data requirements for gross valuation are much more relaxed, as it is necessary only to get the expected (average) generation for the future as opposed to capturing the hour-to-hour variability of wind generation.[1] Credible wind project proposals will have gathered at least a year of on-site data, correlated to longer-term data stations, and converted the wind speed measurements to expected generation. Utility analysts can usually request that data from the project developers, or require such data as part of a formal request for proposals. Where on-site data are not available, other means may be used (detailed methods are presented in Chapter 4), but can be much more complex and error-prone.

3.1.3 Value of renewable energy credits and emissions reductions

Renewable energy credits (RECs—also known as tradable renewable certificates (TRCs)) are certificates of generation from renewable energy generators that convey the right to claim ownership of the environmental attributes of such generation. Generally, one REC is equated to one megawatt-hour of renewable energy generated by a specific renewable generator at some time, or in some time window (e.g. year, quarter, or month). The value of any such right is certainly in the eye of the beholder. One of the main purposes of developing the concept and legal framework for RECs was to allow proponents of renewable energy to place a premium on the generation from renewable resources, and thereby promote more power generated from renewable energy sources.

Some regions have so-called renewable portfolio standards that require a minimum percentage of electric consumption to be produced by certain qualifying renewable energy sources. In many cases the requirements can be satisfied with RECs. A big advantage of allowing RECs to satisfy renewable standards is that the renewable energy can be produced in a region where there is an abundance of the resource, without having to specifically acquire transportation to the targeted demand. A disadvantage is that other, potentially non-renewable generation is serving the region with the requirement. Nevertheless, to the extent that these standards increase the overall generation of renewable

[1] The difference in average generation from hour to hour should not be confused with the hourly variability. For example, the expected wind generation on hour 3:00 in January will not be very different from that of hour 4:00 in January for most wind projects—however, the wind on any particular January day could change quite rapidly from hour 3:00 to hour 4:00, and sometimes (often) not at all. Calculating reserve requirements is all about the potential variability from hour to hour, whereas the gross valuation depends only on the average behavior.

energy, emissions savings will occur somewhere. Geographic restrictions may be enacted so that the renewable generation must come from a region reasonably close, usually electrically interconnected with the target demand area.

Trade in renewable energy credits represents a relatively young and typically not very liquid market (Bird et al., 2007), but this is changing. Europe is ahead of the USA in developing REC markets. The European commitment to meeting greenhouse gas emission goals resulted in establishing minimum REC purchase requirements, fostering a relatively robust market for them. To date, REC trading in the USA has been dominated by voluntary renewable energy programs hosted by utilities, and by individual consumers wanting to promote development of renewable energy. As more and more states enact minimum renewable standards that allow or require compliance through RECs, the market will undoubtedly grow and become more liquid.

For the analyst trying to value wind generation, there are two general approaches to valuing the RECs produced by wind projects. One view of REC value is that it represents the above-market costs of producing renewable generation. In this view, the value of the wind REC is derived by comparing the cost of wind power generation to some proxy resource—often a natural gas fueled generator. The difference in cost, expressed in dollars per megawatt-hour, is taken to be the value of the REC. A potential problem with this valuation is that it does not take into account supply and demand market drivers. For example, regions with a relatively limited wind resource may enact stringent renewable standards that can drive up REC prices if supply cannot keep pace with demand.

REC value can be approached similarly to energy valuation, using the concept of price curves. Unfortunately, commercial vendors have not yet begun supplying REC price curves into the future. Analysts can tap the growing REC broker industry to get current market price quotes, some idea of the local history of REC prices, and brokers' opinions about where REC prices may be headed over time. Clearly, this exercise does not have the kind of precision power planning analysts may hope for, but there may be little alternative until more robust markets for RECs develop.

Perhaps the greatest social benefit associated with wind generation is the reduction in emissions from other power plants that would otherwise be needed to serve power demand. Where direct taxes are levied on pollutants released to the environment, the value of reduced emissions may be straightforward. More typical are 'cap-and-trade' systems in which the total release of certain pollutants within a region is capped, and 'allowances' or 'permits' to release pollutants must be obtained in sufficient quantities to cover the releases. The allowances are sold or auctioned, directly adding to the costs of generators releasing the controlled pollutants. Wind power typically reduces the need for the allowances by reducing the generation from thermal generators. That value needs to be assessed as a direct benefit of the wind generation. Although federal carbon emissions standards are not yet imposed in the USA, at this time there is one mandatory regional system in place in 10 Northeastern States (Regional Greenhouse Gas Initiative—RGGI), and discussions in place for at least one other involving the Western States and several Canadian Provinces (Western Climate Initiative). The regional efforts may be overtaken by federal or international action in the near future to combat the serious threat posed by increasing atmospheric levels of carbon dioxide (IPCC, 2007).

3.1.4 Cost of holding additional reserves due to wind variability and uncertainty

Power system operators ensure there is sufficient ability to increase or decrease generation as needed to accommodate the variations in demand and sudden equipment outages that can occur on the power system. Generally, there are opportunity costs associated with setting aside generating capability for these purposes. For example, to ensure the ability to increase generation on an operating unit, the output must be decreased from its maximum capability. The reduction in generation may be a lost opportunity for a market sale or, conversely, to reduce the generation on a generating unit with higher costs.

There can be costs associated with maintaining the ability to decrease generation as well. For example, at night most power-generating units may be generating at minimum operating levels due to low demand. If there is a need to drop generation further, say to accommodate a sudden increase in wind generation, power system operators may have to take a generating unit off-line completely and start up a smaller or more flexible unit in its place. Many generating units, especially thermal units, can have very high start-up and shutdown costs.

The cost of holding reserves on predominantly hydro-based systems can be almost entirely due to ensuring sufficient ability to *reduce* generation as needed. Hydro projects typically are constructed with more generating capability than is normally needed in order to take advantage of especially high flows that can occur from time to time due to normal weather variations. For such systems, there is usually a large ability to increase generation (at least for many minutes or hours), but limited ability to decrease generation (especially at night) to maintain minimum streamflow levels required for environmental concerns.

Incremental costs of maintaining more generating reserve on many hours can be minimal on both thermal and hydro systems. As just mentioned, the cost of maintaining incremental reserves (ability to increase generation) on hydro systems can be essentially zero except during relatively extreme runoff conditions. So too with a thermal system with sufficient units standing ready to balance increases in demand, any generating capability not needed on any given hour for load is available for another purpose.

Similarly for decremental reserves (ability to decrease generation), there may be very low or zero costs associated with maintaining the needed ability to reduce generation. On many hours there will be generating units operating to meet demand that can be turned down as necessary if the wind should suddenly increase.

This mixture of times when the additional cost of providing additional reserves may be zero, or conversely quite high, makes the cost calculation somewhat complicated. One approach to assessing the costs would be to enumerate the times when the costs are non-zero and estimate the costs of providing reserves at those times—assuming perhaps that a less-optimal mix of generators is needed to meet the total demand, or that an additional generating unit must be energized. For analysts with detailed dispatch models, the necessary reserve requirements are often an input to the model that can be varied to determine the incremental cost of holding the reserves.

The overall cost of holding incremental reserves can be extremely different from utility to utility, depending on the type of resources in the utility portfolio and other factors such as the nature of the existing demand and its size relative to the generating capability of the existing resources. Often the cost of holding reserves will not be very linear. For example, an existing low-efficiency power plant already on the system may be largely unloaded and available to provide considerable reserves with little additional cost—however, as the quantity of reserves surpasses that available from the existing unit, the reserves must come from some other, potentially much more expensive source.

Liquid markets for reserves exist in some places. For such systems, it may be possible to estimate the cost of reserves by looking at historical prices and extrapolating forward. The future of wholesale electric markets may well move toward establishing liquid markets for reserves. Analogous trades occur in other markets as standard call and put contracts. A call contract gives the bearer the right to a certain quantity of commodity at a stated 'strike' price from the seller of the contract. A put contract gives the bearer the right to deliver a certain quantity of the commodity at the stated 'strike' price. Such so-called option contracts are essentially the equivalent of services provided by incremental or decremental reserves. The sale price of the option contract is analogous to the cost of holding the reserves—giving the bearer the right to call on energy or deliver energy. The cost of exercising the reserves is maintained in the strike price. The price paid for an option contract is called the premium. In a liquid wholesale electric market where option contracts are prevalent, the cost of holding reserves would be the premium price of the needed contracts. Some electric power markets have not yet quite evolved to this point, leaving analysts to calculate the cost of holding reserves from knowledge of their own systems and how they behave.

Not only is the cost of holding reserves nonlinearly dependent on the amount of needed reserves; it is also dependent on the type of reserves. Generating units vary in their ability to respond to the need to increase or decrease generation. The two main parameters have to do with how quickly a generating unit can begin to respond to a signal to change output level and how fast the level can change once it has begun. Thermal units already energized and synchronized with the grid ('spinning') can begin generating more energy in a matter of seconds, whereas a unit that has not been energized may take anywhere from 10 minutes to several hours to begin generating power.

Some reserves are normally held to respond to sudden equipment outages, called 'contingency' reserves in the USA. Usually some fraction of that reserve must be made up of fast responding (less than 10 minutes), with the remainder coming from slower reacting (usually within 30 minutes) units. Other reserves are needed to respond to the normal variability of the load. These are often similarly broken down into faster-acting reserves (often referred to as 'regulating' reserves in the USA) and slower-acting (often referred to as 'following' reserves in the USA) reserves. The faster-responding regulating reserves are typically more expensive to provide, necessitating more flexible generating units. The slower, following reserves are typically less expensive services to provide. It is therefore important to distinguish the types of reserve needed and available in the costing analysis.

Wind generally represents a relatively small need for additional regulating reserve—perhaps less than 1% or so of the nameplate wind installed. For example, perhaps 10 megawatts of incremental regulating reserve needed for more than 1000 megawatts of installed wind. The following reserve requirement is typically much higher—anywhere from say 5% to 20% of nameplate wind installed, depending on the amount of wind installed relative to the size of the existing system.

Another component that can potentially add to costs is ramp rate. Ramp rate is the rate, usually in megawatts per minute, that generation can change—though usually the limitation is on the needed rate of increase, not decrease. Ramp rate has less often proven to be an issue for power systems with wind on them, but the analyst needs to be aware of the potential for a problem, and should do at least some rough estimates to ensure there is sufficient ability to match the needed rate of increase or decrease in generation as the wind changes.

Some of the organized trading markets, such as the Nordic Pool in Scandinavia (IEA, 2009), use energy markets (e.g. trading in 10- or 15-minute increments) to provide much of the needed balancing services. The effects of market structures on the valuation of wind generation is explored in more detail in Chapter 11.

3.1.5 Effects on reserve generation operating costs

The previous section covered the cost of maintaining the ability to increase or decrease generation as wind comes up or goes down within an operating period. That cost is largely based on the opportunity cost associated with maintaining capability in case it is needed to balance the wind—a cost that may be incurred even if the wind happens to be holding steady, or not blowing at all at any particular moment. There is a separate and distinct cost incurred when generating units are called upon to either increase or decrease generation in response to the joint effects of wind and load. The incremental cost of operating reserve units due to added wind generation is the subject of this section.

Figure 3.3 shows a typical gas turbine heat rate curve, illustrating the relationship between generating unit efficiency and operating level. Most generating units are optimized to operate at maximum efficiency near the maximum output level. A unit that must maintain the ability to increase generation must be operated below the maximum generating capability, and hence at a lower efficiency than otherwise would be experienced. For example, if the generator depicted in Figure 3.3 were operated at 100% of rated output, it would generate at an efficiency of 43.8%, consuming 7800 BTUs of fuel to produce each kWh of electric energy. If the generator is operated at 80% of rated output, it would drop to 42.7% efficiency, thereby consuming an additional 100 BTUs for each kWh produced. This drop in efficiency could be attributed to the opportunity cost identified in the previous section, but there is another effect that needs to be considered.

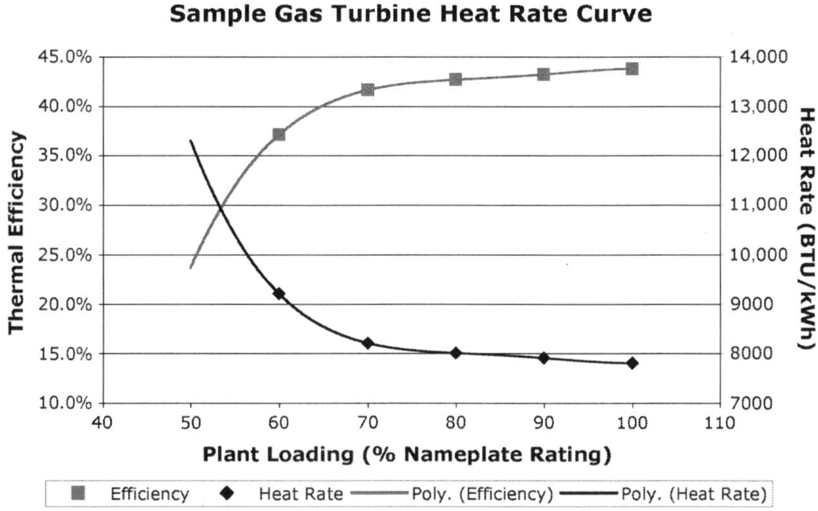

Sample Gas Turbine Heat Rate Curve

FIGURE 3.3 Sample gas turbine operating characteristics, showing the rapid deterioration of generating efficiency as the turbine is operated below nameplate rating.

It is a relatively simple calculation to determine the increased cost from operating a turbine at a lower *constant* level due to holding more reserves on the unit. However, because the curve is not linear, operating the unit at both higher and lower levels than the average point incurs additional reductions in efficiency. As an example, consider the turbine in Figure 3.3 operating at an average of 80% of nameplate, consuming 7998 BTUs for every kWh generated. If the unit were operated a third of the time at 80%, a third at 90%, and a third at 70% (still averaging 80% overall), the average heat rate is 8016 BTU/kWh. The difference, 18 BTU/kWh, is due strictly to the operation of the unit over the range of its heat rate curve. The effect increases with the overall range over which the unit is operated and the average level—the nonlinearity in the heat rate curve often becomes severe at lower generation levels.

Another important effect is the number of starts and stops that units undergo over the year. Unit starts often entail additional costs, at least some of which stem from operating the unit at the severely inefficient levels (albeit for relatively short periods) in the lower end of Figure 3.3. Thermal units may have to be started more times, or more of them started to cover the increased need for reserves. These costs can be significant contributors to the cost of integrating wind.

Finally, there is likely additional wear and tear on generating equipment having to ramp generating levels up and down more frequently. This is especially so for large thermal units with boilers and steam generators, where massive pieces of equipment retain large amounts of heat energy in their thermal mass. Calculating these additional costs can be a daunting task and often is done either through third-party estimates, or through interviews with plant operators. Identifying additional wear and tear has been an inexact science for the most part.

3.1.6 Balance of system and market trading costs

Just as incremental demand has the potential to change the fuel consumption at power stations throughout a power system, so does the addition of an incremental generating unit potentially affect other generators. In general, adding wind generation to a power system decreases overall fuel consumption, and reduces fuel costs that would be experienced absent the wind. Wind generators have very low (potentially zero or even negative) operating costs, so generators with higher operating (especially fuel) costs will tend to operate at lower levels. The need to hold reserve generation to accommodate the variability and uncertainty in wind generation is a complicating fact, as pointed out above. Savings in fuel costs are generally covered under determining gross value of the energy, but the effect of large amounts of low variable-cost wind generation can alter market prices and the marginal cost of generation across a power system.

It may turn out that the optimum resource displacement occurs in power plants located in adjacent power systems. That, and the basic variability and uncertainty in wind generation, tends to increase the frequency and extent of trading in day- and hour-ahead markets. A thorough valuation analysis allows for such trading and will include market transaction costs that may include sufficient transmission wheeling costs to reach the appropriate markets.

Costs associated with providing balancing services and various penalties or wind-specific costs may be levied directly by the power system into which a wind project interconnects. If such charges exist, they will make up a part of the valuation analysis. However, the costs incurred by providing balancing services may not be known by the local utility or transmission system operator. The following chapters should be helpful in identifying and calculating such costs directly.

3.2 SUMMARY

This chapter outlined the general components that any thorough analysis of wind power economics will address. Realistically, not every last detail is encompassed in every wind integration study. It is reasonable to identify the issues of major concern to a particular utility. Hydro utilities may focus more on operational limitations and the cost of providing down-regulation reserves, while a coal-based utility may be more concerned about ramp rates, unit starts, and regulating (fast-responding) reserves.

Not every aspect of wind is new, or specific to a wind valuation analysis. The cost of the power plant, financing costs, and value of the resulting generation are issues in common with other technology valuations. Evaluating the value brought by RECs may be a new concept for some utilities, but the concept of estimating the future value of a commodity is similar to estimating market prices for electricity. Specific to wind are the costs of holding incremental balancing reserves, and the cost of dispatching those reserves to meet the increased variability and uncertainty that are somewhat peculiar to the wind resource.

The purpose of this chapter has been to outline the work to be done, without delving into specific details. It may be apparent from this discussion that the behavior of wind is of the utmost import to calculating these costs. Analysts must develop wind generation data from which certain important statistics can be developed. For example, one cannot determine the amount of reserve needed (or the type) without a data set from which to measure wind's variability. Similarly, the frequency of dispatch of resources to balance changes in wind output necessitate a detailed understanding of both the magnitude and frequency of the variations brought by wind.

The next step, really the first and often most challenging step in a wind analysis, is to develop an estimate of wind generation data. Developing a reliable representation of wind output is the subject of the next chapter. Following that will be a discussion of how the wind data can be characterized for use in dispatch models, and Chapter 6 covers how to determine reserve requirements from the wind data set.

REFERENCES

Bird, Dagher, & Swezey. (2007). *Green power marketing in the United States: A status report* (10th ed.). National Renewable Energy Laboratory. NREL/TP-670-42502.

Bolinger, M., & Wiser, R. (2005). *Utility integrated resource planning: An emerging driver of new renewable generation in the western United States*. Lawrence Berkeley National Laboratory, LBNL-59239. Retrieved from: <http://www.escholarship.org/uc/item/1sf3k4xv>

Intergovernmental Panel on Climate Change (IPCC). (2007). *Fourth assessment report. Climate change 2007: Impacts, adaptation, and vulnerability*. Cambridge University Press.

International Energy Agency (IEA) (2009). *Task 25: Design and operation of power systems with large amounts of wind power*. Final Report, IEA Wind Task 25. VTT.

Developing Useful Wind Generation Data

The breeze at dawn has secrets to tell you.

Rumi, thirteenth century Persian poet

One of the most difficult challenges in performing wind valuation studies is developing accurate wind project generation data to use in the analysis. Long-term averages of wind generation on an hourly basis for each month, or at least each season, are necessary to determine the gross value of the wind generation and are commonly produced in the process of wind project site assessment. Producing data sets that adequately characterize the variability and uncertainty of wind plant output for the purpose of determining reserve requirements is more demanding.[1] Developing good data sets is complicated by the fact that most studies are prospective in nature—requiring data for wind projects that have not yet been constructed at sites for which limited data may be available.

Relatively standard techniques can be employed to develop wind speed data, but techniques for converting wind speed data into wind generation are at an earlier stage of development and contribute to the uncertainty of the final product. Some of the approaches analysts have taken are addressed in this chapter.

There are several avenues open to analysts in developing wind data for reserve analysis, and the chosen approach will at least partly depend on the available data. Performing a study of existing wind power facilities for which historical generation data are available (i.e. retrospective analysis) is the most straightforward problem to undertake. The opposite extreme is analyzing projects for which no on-site data are available and whose very geographic locations may not be specifically known.

[1] In addition to basic wind generation data, determining reserve requirements necessitates an evaluation of wind generation forecast errors. Thus it is equally important to produce synthetic wind power forecasts. Synthetic forecasts are covered in Chapter 7.

Valuing Wind Generation on Integrated Power Systems. DOI: 10.1016/B978-0-8155-2047-4.10004-3

For many prospective analyses, wind forecasting service and information providers are employed to produce credible historical wind speed sequences across regions of interest using numerical weather prediction models.[2] Additional processing is necessary to convert historical wind speed data to generation levels for specific project sites. Analysts need a working knowledge of how these data are produced, especially the conversion from wind speeds to wind project output.

The main alternative approaches to developing wind project data include:

- Direct use of wind generation from existing facilities in a retrospective study.
- Using data from existing wind projects to estimate output of prospective projects at other (potentially nearby) projects.
- Using available on-site wind speed measurements to estimate project output.
- Using weather model wind speed data to estimate prospective wind project output.

Methods for developing wind generation data from prospective plants are necessarily approximate. Unfortunately, the reserve margin requirements, and hence integration costs, are sensitive to small errors in the analysis. It is therefore important that certain statistical properties of the modeled data are reasonably well estimated. The level of reserve margins necessary to maintain power system reliability standards involves determining extremes of the changes in aggregate wind generation from one time period (e.g. 10 minutes or an hour) to another. Accurately setting reserve margins requires accurately assessing the magnitude of the relatively low probability events. The accuracy of wind integration cost studies has sometimes hinged, or perhaps become unhinged, on the accuracy of the estimated wind plant output properties, and it is very important to do a credible job of developing the data. Unfortunately, the analytical methods are more art than science at this time. This chapter will focus on methods for developing wind generation data and examine some of the properties of the results to avoid obvious errors.

4.1 SENSITIVITY OF STATISTICS TO SCALING

Wind integration costs are primarily associated with the incremental need to hold and dispatch reserve units to counter rapid and unexpected changes in wind output together with variations and uncertainty of demand. It is therefore important in developing wind project output

[2] As of early 2009, the US National Renewable Energy Laboratory maintains a database of wind speed data for much of the Western US based on numerical weather prediction model runs for years 2004–6 for 2-km grid segments (see http://wind.nrel.gov/public/WWIS/Wind_Data/Terms_of_use.htm). The NREL's data set includes approximated generation output of 30 MW wind projects at selected sites.

estimates to be careful to preserve the aggregate variability of wind projects from one time period to the next. An important data feature to capture is the extent to which geographically separate wind projects change generation levels in concert with one another—exacerbating the aggregate variability.

As Figure 2.6 suggests, the most significant variations in wind output occur over periods of many minutes to several hours—large-scale changes in aggregate output do not occur over periods of a minute or less for multiple-wind-turbine generating stations (IEA, 2007). A common statistic on which reserve margins may be based is the 95th percentile of 10-minute on 10-minute changes, to remain in compliance with the North American Electric Reliability Corporation's Control Performance Standard 2 (CPS 2).[3] The granularity of time increments for which wind data may be available generally ranges from a second to an hour.

In the case of an expansion of an existing wind facility, it is tempting to simply 'scale up' the output of the existing facility. For example, the output of an existing 25-MW wind turbine project might be multiplied by 4 to represent the output of a 100-MW expansion. While some characteristics may be acceptably reproduced under this operation (e.g. maximum generating capability), the variability of the output is not one of them (Holttinen & Hirvonen, 2005). The variability of wind project output, as a fraction of total nameplate capacity, declines with the number of turbines and geographic dispersion of those turbines. This is suggested by the declining correlations over distance illustrated in Figure 2.6. One example of this effect is shown in Table 4.1, in which the variability of 10-minute data is examined for three adjacent wind projects. Table 4.1 summarizes statistics for 10-minute production data expressed as a fraction of the installed nameplate capability. For the 16-turbine wind project, 95% of the 10-minute changes are less than or equal to 6.8% of the nameplate. For the 235-turbine adjacent wind project, the 10-minute changes are less than or equal to just 3.9% of nameplate capability. Negative figures in the 2.5 percentile row represent reductions in output from one 10-minute period to the next.

For the sample data summarized in Table 4.1, if the variability of the 16-turbine data were simply multiplied by a constant factor to represent the variability of the 235-turbine field, it would overstate the reserve contribution from the larger field by nearly a factor of 2.[4] Similarly, starting with

[3] NERC CPS 2 requires balancing areas to maintain 10-minute average area control error (ACE) within certain bounds 90% of the time, measured monthly. To ensure compliance and avoid penalties, balancing areas generally target a higher level of confidence, hence the 95% figure used in the text. See http://www.nerc.com/files/BAL-001-0a.pdf

[4] The reserve requirements are determined from the combination of all variability in the balancing area—including the net effects of all wind projects, loads, and other resource variability (e.g. run-of-river hydro generation).

TABLE 4.1 Statistics on 10-minute Production Data as Percentage of Nameplate Generation

	16-Turbine Deltas	50-Turbine Deltas	235-Turbine Deltas
Standard deviation	5.0%	4.1%	2.7%
97.5 percentile	11.3%	9.7%	6.1%
95 percentile	6.8%	5.9%	3.9%
5 percentile	−6.7%	−6.0%	−3.7%
2.5 percentile	−11.0%	−8.8%	−5.8%

a larger wind project and scaling it down (multiplying by a factor less than 1.0) would understate the variability brought by the smaller wind generating facility. This example illustrates the need to use extreme caution in using multiplicative scaling factors. Figure 4.1 shows the reduction in hourly variability as the effective area of the wind projects increases.

One method proposed is to use weighted moving averages of the time series to smooth out the time variations of the smaller projects to represent the larger ones (Nørgård & Holttinen, 2004). For example, if a time series representing the output of an existing project is $g_1, g_2, g_3 \ldots$, the new series can be constructed to represent the output of a sister project $G_1, G_2, G_3 \ldots$ by applying fixed weights w_{-1}, w_0, w_1:

$$G_i = w_{-1}g_{i-1} + w_0 g_i + w_1 g_{i+1} \tag{4.1}$$

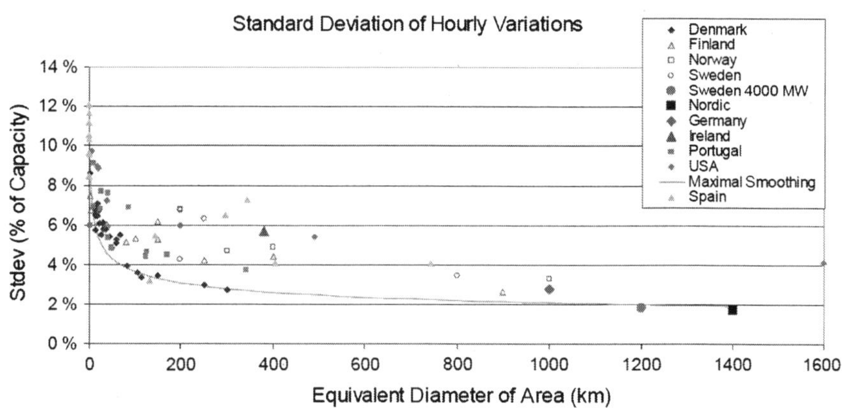

FIGURE 4.1 The reduction in the standard deviation of hour-to-hour wind generation as the effective area of the wind projects increases. *Source: Design and operation of power systems with large amounts of wind power. Final report, IEA Wind Task 25, Phase one, 2006–2008, Fig. 7, sect. 2, <http://www.vtt.fi/inf/pdf/tiedotteet/2009/T2493.pdf>.*

The weights should sum to 1.0 and are chosen to smooth the variability to reproduce the behavior depicted in Figure 2.6 and Table 4.1. More than three weights can be chosen as necessary to achieve the desired level of smoothing.

There is a physical interpretation of the weighting methodology that reappears in other techniques described in this chapter. The idea is that changes in wind speed propagate at the average wind speed in the prevailing wind direction. Geographically larger wind-turbine fields will intercept a larger temporal window of wind speeds. Changes of wind speeds in time are translated to changes of wind speed across a region. This concept is illustrated in Figure 4.2.

4.1.1 Scaling to nearby wind projects

The technique suggested by equation (4.1) may be useful in generating data for nearby expansion plants, and has also been used to develop data for more distant projects by inserting a time shift representing the travel time from the point at which the data were collected to the target project area. There are some important limitations to this theory to consider.

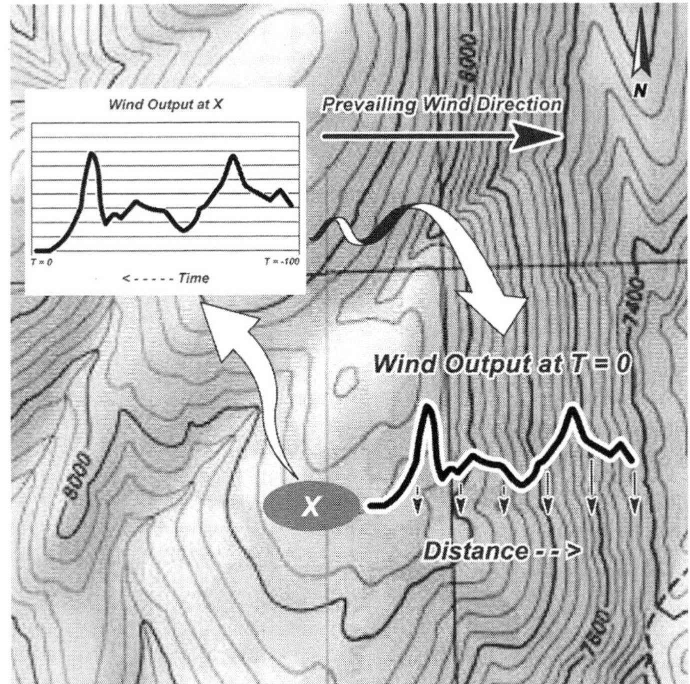

FIGURE 4.2 The concept that changes in wind output or wind speed in time measured at one point can be interpreted as representing the distribution of wind in an adjacent region as the changes in wind output propagate across the wind-turbine field.

One such limitation is that the physical interpretation suggests only the distribution of wind speeds in the direction of the prevailing wind, not in the orthogonal direction. This is especially limiting for several reasons. First, wind turbines are usually arrayed in lines perpendicular to the prevailing wind direction and not parallel to it in order to minimize the wind shadowing effect. Second, significant wind events often arise from the passage of weather fronts where the direction of the wind is often perpendicular to the movement of the front and the speed of the advancing front may not be related to the wind speeds themselves. Another basic issue with the method is that much of the energy in surface winds comes from turbulent mixing in the vertical direction, a phenomenon completely overlooked in this approach. A further transformation may be necessary if the quality of the wind resource (i.e. average wind speed) at the target site is expected to be different from the site from which the data were taken.

Inserting a simple time shift to an existing data set is insufficient to represent the output of a neighboring wind project. At a minimum, an additional random factor must be added to the data to accurately capture the diversity of wind projects (Xie & Billinton, 2009). Determining the level of randomness to add into the scaling function depends on the level of correlation expected between the sites, and differences in geography and climatology between the sites. In the absence of at-site measurements, correlations among existing sites in the region or correlations in weather model data might be used.

4.2 CONVERTING WIND SPEED TO WIND OUTPUT

Wind-turbine power curves form the basis for translating wind speed measurements to power output. Wind-turbine manufacturers produce wind-turbine power curves to characterize the amount of power generated at various wind speeds for their turbines. A generalized power curve is illustrated in Figure 4.3. The cut-in speed is defined as the wind speed at which the wind turbine begins to generate power. For a turbine characterized by Figure 4.3, that wind speed is approximately 4 meters per second. Wind turbines are designed to shed the wind and shut down to protect them during excessively high wind conditions. This occurs at approximately 24 meters per second in the illustration. Note that power curves from different manufacturers and with different turbine configurations (e.g. turbine blade lengths) will have different power curves.

From the foregoing, calculating wind generation data would seem to consist of taking wind speed data, potentially adjusting for hub height and applying the appropriate power curve. Several complicating factors make an accurate calculation somewhat more involved (Potter et al., 2007). Power curves are idealized representations of wind-turbine output

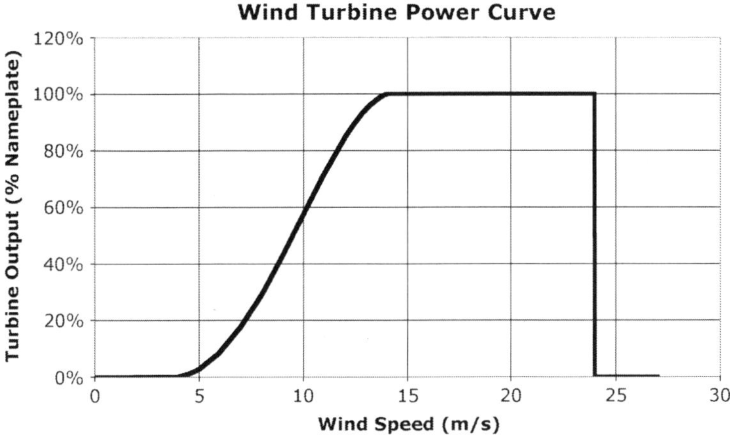

FIGURE 4.3 Illustrative wind-turbine power curve.

for wind speeds. In practice, the direction of the wind is an important factor as nearby turbines or geographical features can have wake ('shadowing') effects on generation levels. Even accounting for wind speed, the observed data tend to scatter around the power curve, with data points scattered about the power curve. Factors other than wind speed and direction clearly affect the observed wind generation from real-world wind turbines. Nevertheless, the power curve is an essential building block in developing a model of wind generation from wind speeds.

4.2.1 Adjusting wind speed measurements to hub height

Wind speed measurements taken at the hub height of the wind turbine can be applied directly to the power curve to determine the output of the turbine. Weather models generally provide wind speeds at an array of elevations above ground, but data from on-site measurements may be taken at heights somewhat lower than the hub height of modern turbines. Wind speeds tend to rise with elevation and it is necessary to adjust wind speed measurements taken at lower elevations to turbine hub heights (80–90 meters with current technology). The relationship between elevations and wind speeds is usually taken to be:

$$\left(\frac{h_1}{h_2}\right)^{\alpha} = \frac{v_1}{v_2} \tag{4.2}$$

where h_1 and h_2 are wind speed measurement elevations, and v_1 and v_2 are the corresponding wind speeds. The exponent α is the 'wind shear' coefficient for the local area. The wind shear is estimated by taking simultaneous wind speed measurements from a meteorological tower at different heights and using equation (4.2) to solve for α. The equation can then be applied again to adjust the direct measurements to hub height. Wind shear coefficients range from 0.05 to 0.25 over land, with typical values ranging from 0.14 to 0.20 (Ray et al., 2006).

4.2.2 Multi-turbine power curve equivalent

Wind speed measurements are generally taken at one or a few points on prospective wind project sites. Applying a point estimate of wind speed to a power curve representing the turbine model to be installed results in the expected output of a single wind turbine located at the specific location at which the wind speed data were collected. It is important that the data be collected at a location that is expected to be representative of the average wind speed over the entire turbine field. Collecting the data from the best part of a turbine field will overstate the performance of a large collection of turbines located around the site. In any case, the measured values will be a point estimate for wind speeds around the turbine site. At best, the values will represent the average wind speed for a wind project site. Because power curves are not linear, the output of multiple wind turbines located around the site is not the same as the average wind speed experienced by the turbines applied to the single-turbine power curve. A couple of simple examples will illustrate this point.

Consider a wind-turbine project consisting of just three wind turbines characterized by the power curve illustrated in Figure 4.3 and having a nameplate rating of 2 MW each. At some point, the wind speeds experienced by the three turbines average 4 m/s, specifically realized as 3, 4, and 5 m/s at the individual turbines. Applying the 4 m/s average wind speed to the power curve results in an output of 0 MW. However, the turbine experiencing 5 m/s wind speeds is actually generating at a rate of approximately 6 kW (3% of nameplate).

One way of producing the output of multiple turbines from a single data point is to construct a power curve that represents the combined effects of multiple wind turbines (Nørgård & Holttinen, 2004). If a distribution of wind speeds for the multiple turbine site can be associated with each measured value, they can be individually applied to the single turbine power curve, and then summed to determine the effective power curve for the combined set of turbines. Taking the previous example further, and assuming that the three turbines are adequately modeled as a distribution in which one turbine takes on the expected value and the other two turbines are at plus and minus 25% of the wind speed, a revised power curve can be constructed from the one in Figure 4.3. Such an example is a bit extreme—more often a continuous distribution of wind speeds across the wind turbine site is assumed. The effect of assuming a normal distribution of wind speed on the power curve in Figure 4.3 is shown in Figure 4.4. A primary effect of multiple wind turbines is to smooth the effective power curve.

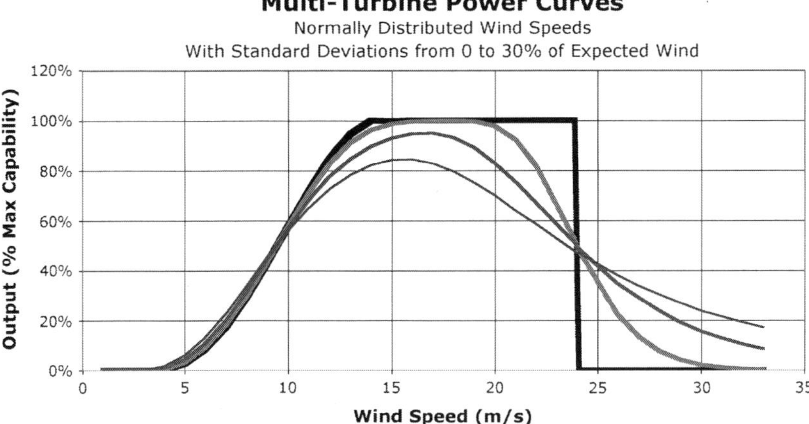

FIGURE 4.4 The effect of multiple turbines on the equivalent power curve.

The amount of power curve smoothing depends on the breadth of the distribution of wind speeds across the turbine site. That distribution is dependent on two primary factors: the size of the site and the inherent wind turbulence intensity at the site. Turbulence intensity I is defined as the standard deviation of wind speeds over some time period, divided by the average wind speed over that period (Burton et al., 2001):

$$I = \frac{\sigma}{\overline{V}} \tag{4.3}$$

where $\overline{V}$ is the average wind speed and σ is the standard deviation of wind speed. Direct measurements of turbulence intensity for specific sites, at different average wind speeds, in different seasons, and hours of the day would be useful in developing appropriate wind speed distributions to use in developing the equivalent power curve.[5] A detailed model of individual wind sites may be warranted; however, analysts' lack of time and resources may require using whatever information is available to make as educated an estimate as possible in this regard.

Both project size and wind turbulence intensity contribute to the variability of the wind speeds across the field of wind turbines in a project. Figure 4.5 shows a relationship (Nørgård & Holttinen, 2004) developed to show the variability of wind speeds as a function of turbulence intensity and project size. The figure presents normalized standard deviations (standard deviation/average wind speed) for wind

[5] Note that this implies a family of power curves that vary depending on wind direction, season, and hour of the day. The author is unaware of any implementation of such a scheme.

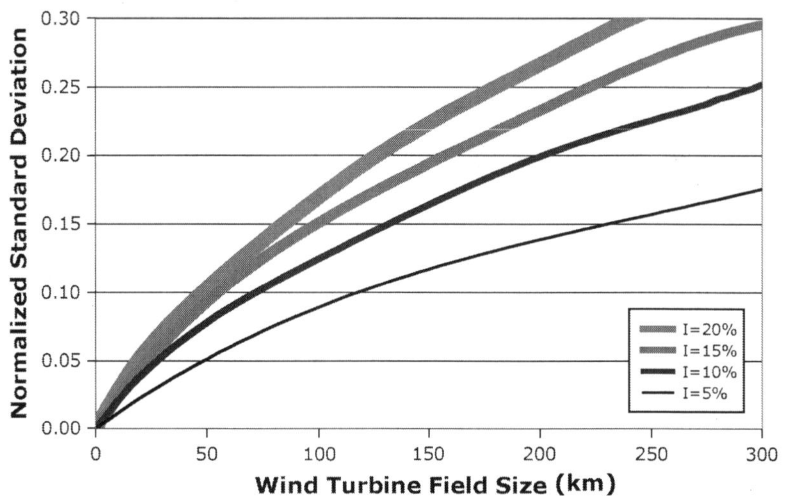

FIGURE 4.5 Increasing variability of wind speeds with distance (effective project size) for different levels of wind turbulence intensity. *Adapted from Nørgård & Holttinen (2004).*

turbines over distances and turbulence intensities. Note that the shape of the distribution is also important. A study produced for NorthWestern Energy of Montana (USA) wind variability (Genivar, 2008) assumed a blanket 10% wind turbulence intensity and normally distributed wind speeds. The Genivar study tested different levels of wind turbulence intensity and found little effect on the resulting estimate of wind generation variability.

4.2.3 Block-averaged wind speeds

Before applying wind speed measurements to the multi-turbine equivalent power curve, it may be necessary to adjust the data to extrapolate from the point estimate to the average wind speed (or a distribution of wind speeds) over the project site. Once again invoking the assumption that changes in wind speed approximately propagate at the average wind speed, the measured data should be averaged over a time wind representing the project site size, as discussed in Section 4.2.2. In the example of Section 4.2.2, the project size of 1 km in a 10 m/s average wind speed results in a 100-second time window representing the project site size. A 1-km project might therefore be represented by taking a moving average of the available data over 100 seconds of data.

This procedure assumes that the data are available on a timescale less than the project timescale, which is often not the case. In this example, if the wind speed measurements were only available on a 10-minute basis, no averaging would be necessary because the data are already averaged over 600 seconds. This method is further complicated by the fact that the length of the moving average presumably changes with wind speed. However, for very large projects, or in using a single wind speed measurement to represent several dispersed projects, the moving average may be useful.

4.3 USING WEATHER MODEL DATA

An effective use of wind speed data generated from numerical weather models is to extend on-site data measurements taken over a limited time period (a minimum of 1 year is desirable) to longer historical records. Weather model output can be calibrated to the on-site measurements to improve the accuracy of the representation for other time periods. Producing data for other time periods enables development of wind generation for many projects—both existing and prospective—over identical historical time periods (e.g. the same historical year or years). Analyzing data over consistent time periods is necessary to determining reserve requirements, a central determinant of wind integration costs.

When on-site data measurements are unavailable, it may be necessary to resort to using wind speed data produced by numerical weather prediction models directly. Such data may have already been produced for specific historical time periods covering the desired sites, or can be produced for that purpose by firms such as those listed in Appendix A. In general, it is preferable to rely on direct wind speed measurements than on the output of weather models. As just noted, model accuracy is greatly enhanced when calibrated against at least 1 year of direct measurement data at, or near the intended site. One reason for this is that important topographic features can be missed by these models that divide the surface of the earth into grids that are rarely finer than 2 km on a side. Nevertheless, studies where no on-site data are available may necessarily rely on weather model data alone.

Data produced by the models typically represent wind speed in the center of each grid point at typically 10-minute intervals.[6] These wind speeds must be converted to aggregate wind generation over multiple wind turbines of a wind project. Each turbine will experience a wind speed different from the average and it will be necessary to produce a set of wind speeds around the model estimate (Potter et al., 2007). Given current computing technology, grid sizes used in weather models are at least 1 km square. Topographical features that affect wind speeds are potentially significantly smaller than 1 km in extent. As a result, wind speeds reported

[6] Although the data are typically reported out at 10-minute intervals, the models internally calculate variables more frequently. A common rule is to have the model calculate variables at intervals of $3x$, where x represents the grid size in km and $3x$ is the interval time in seconds.

by weather models tend to be less volatile in time and more highly correlated in space than directly measured wind speeds.

More accurately reflecting observed wind speeds necessitates adding back in stochastic error terms to reduce the overly correlated and smoothed data from these models similar to that described in Section 4.1. It has been suggested (Smith et al., 2007) that the smoothing introduced by weather model wind speed estimates is approximately equivalent to the smoothing across wind turbine fields of 30–40 MW in size for each reported data cell (geographic location). However, this approach does not appear to address over-correlation between and among nearby data cells. Accounting for both the over-smoothing in the models and the stochastic response of wind-turbine generators to wind speeds requires adding uncertainty back into the model results. One method proposed (Potter et al., 2007) uses historically observed distributions of generation levels for a given wind speed to derive a more realistic level of variability. However, care must be taken to preserve the time correlation of wind speeds as well.

Ultimately the application of weather models to produce data wind generation data sets holds the greatest promise for developing representative data. At this point, however, data directly from the models need to be adjusted using methods that have not yet become standardized. Analysts must be vigilant and review the behavior of the resulting wind generation patterns against other available information, such as observed wind speed/generation relationships in the region under study and the correlations illustrated in Figure 2.6.

4.4 SUMMARY

Gathering a reliable data set representing the output of the modeled wind projects is likely the most taxing task facing the analyst in determining wind integration costs. It is vitally important to accurately assess both the variability and the extent to which the output of wind projects are correlated with one another without either over- or understating the effects. The best advice is to do the best job possible with the information and resources at hand. At minimum, the data set should have been developed such that:

- The data represent the output of the collection of wind turbines being modeled, over an historical time period for which power demand data are available.
- If the data are developed from nearby wind project output, care is taken not to scale by a multiplicative factor without further taking into account stochastic differences (i.e. not 100% correlation) between the projects, and the relative sizes of the projects.
- If the power generation is developed from wind speed data measurements, the effects of varying wind speeds should be taken into account (e.g., by developing a multiple-turbine power curve equivalent), and

any necessary smoothing due to project size should be taken into account.

- If power generation data are developed directly from numerical weather models, care should be taken not to overstate the extent to which adjacent projects will appear to be correlated in the unadjusted model output.

A related issue explored more fully in Chapters 7 and 8 is the need for synthetic forecasts of the wind generation data. Synthetic forecasts embody the uncertainty inherent in wind forecasting necessary to assessing costs associated with that uncertainty. Chapter 6 shows how incremental reserve requirements are dependent on forecast accuracy.

REFERENCES

Burton, T., Sharpe, D., Jenkins, N., et al. (2001). The wind resource wind energy handbook. *Wind energy handbook*. Wiley. p. 17.

Genivar (2008). *Montana Wind Power Variability Study*. NorthWestern Energy. Available at: <www.uwig.org/NWE-MontanaWindPowerVariabilityStudyFINAL.pdf>

Holttinen, H., & Hirvonen, R. (2005). Power system requirements for wind power. In T. Ackermann (Ed.), *Wind power in power systems* (pp. 143–167). Wiley.

International Energy Agency (IEA) (2007). *Task 25: Design and operation of power systems with large amounts of wind power: State of the art report*. VTT.

Nørgård, P., & Holttinen, H. (2004). A multi-turbine power curve. *Proceedings of Nordic wind power conference NWPC'04*.

Potter, C., Gil, W., & McCaa, J. (2007). *Wind power data for grid integration studies*. IEEE Power Engineering Society General Meeting.

Ray, M. L., Rogers, A. L., & McGowan, J. G. (2006). Analysis of wind shear models and trends in different terrains. *AWEA windpower conference 2006*.

Smith, J. C., Parsons, B., Acker, T., Milligan, M., Zavadil, R., Scheurger, M., & Demeo, E. (2007). Utility wind integration and operating impact state of the art. *IEEE Transactions on Power Systems, 22*(3), August.

Xie, K., & Billinton, R. (2009). Considering wind speed correlation of WECS in reliability evaluation using the time-shifting technique. *Electric Power Systems Research, 79*, 687–693.

Chapter five

Representing Wind in Economic Dispatch Models

The same wind that blows one ship into port may blow another off shore.

Christian Nevell Bovee, 1820–1904, American author and lawyer

Studies of the value of power systems and components are commonly performed with complex computer models known as economic dispatch models. Economic dispatch models vary in their detail and complexity, but the main features include a representation of power system demand through time and a set of generating resources for meeting that demand. The cost of generating power from the generators is an important part of the representation in that the models simulate the economic operation (e.g. least cost) of the generators to supply the demand. The representation of generators typically includes constraints such as maximum and minimum generating levels, maximum rate of change of output, as well as environmental constraints on hours or levels of operation, etc. Other features such as access to market points for buying and selling power, or limits on the transmission of power to markets, load centers, and generators are also common. The focus of this chapter is on how generators are represented in these models, and in particular how wind generators may be represented in models that are not generally designed to accommodate the specific characteristics of wind generators.

Dispatch models generally allow data entry for different generating resource categories that may include thermal plant (coal, combustion turbines, nuclear, other boilers, etc.), hydroelectric, pump storage, and contract purchases. Notably missing is any category for wind. The author is unaware of any commercial power system dispatch model that has a specific category for wind, or logic dedicated to appropriately modeling wind generation. Without such built-in features, analysts are left to improvise somewhat.

Valuing Wind Generation on Integrated Power Systems. DOI: 10.1016/B978-0-8155-2047-4.10005-5

By far the most common approach is to represent wind generation as a fixed time series of generation levels, either as a zero-variable cost resource or directly as a load reduction. This approach has the virtue of simplicity and ease of implementation, not requiring special logic or changes to any existing dispatch model that may already be available to the analyst. It is important to keep in mind both the implications and limitations of this approach.

5.1 IDEAL REPRESENTATION OF WIND GENERATORS IN DISPATCH MODELS

Before delving into the approximations that are likely necessary to shoehorn wind projects into existing dispatch models, it may be useful to describe how wind generation might ideally be represented. An ideal dispatch model would have several specialized features:

1. **Develop available wind generation data.** This feature would develop time series of available wind data with options to pick specific historical periods to model, or conversely to develop synthetic data to represent wind generators being modeled. Synthetic data would need to: preserve the auto-correlations expected through time at relevant time scales; the correlations among various projects; the average energy (i.e. capacity factor) of the wind generators over months of the year and hours of the day; and the relationship between power system demand and wind generation—specifically the tendency for large high-pressure systems to depress wind generation at times of load extremes. Data should be available at a variety of timescales from 1 minute to several hours.
2. **Model wind generator flexibility.** Modern wind generators have significant flexibility that should be modeled: they can be operated to limit their ramp rates (rate of increase of generation—MW/min), or to produce a fixed fraction of the available wind generation on average while providing regulating reserve, or limit their maximum output (e.g. when all units can produce at maximum it may be desirable to generate at a lower level to reduce the potential for rapid ramp-down). Further, it would be useful to impose limits only under selected conditions—for example, limitations might only be imposed when large-scale weather fronts are forecast to approach.
3. **Model dispatch costs.** Although the fuel is free, wind generators may face opportunity costs when the output is reduced. It is important to include the relevant economics, which may include loss of feed-in tariffs, contract sales prices, revenues associated with renewable energy credits, and loss of production tax benefits.
4. **Model wind and load schedules.** Dispatch models employ relatively sophisticated and typically proprietary unit-commitment logic. This logic looks ahead in time to determine which generating units to start

up or to shut down due to economic and load conditions. The timeframe for these decisions varies with the characteristics of the generators. Some generators may take many hours or even days to start up. An ideal wind model would allow for unit commitment and dispatch to be dependent on wind generation and load schedules (i.e. forecasts). Such forecasts would become more accurate as time approaches the operating period. At a minimum the model would allow for, or create, wind schedules a day in advance and an hour in advance.

Models approaching these ideal features above are unknown to the author at this time. The most lamentable shortcoming is in the realm of power generation data for wind projects. Some progress is being made, but the problem is relatively complex and may not be satisfactorily solved for some time. Taking appropriate account of wind generator flexibility can be expected to become a more significant issue as wind generation becomes a significant part of power systems. Model vendors will likely respond over time by installing more modeling capability for wind generation.

5.2 FIXED TIME SERIES IN FORWARD- AND BACKWARD-LOOKING ANALYSES

Economic dispatch models simulate the operation of a power system over either an historical period of interest or a future period, and in some cases a hybrid of the two. For example, a study may be designed to capture the future value of a change in the efficiency of a hydro plant efficiency upgrade. For such a study, it would be important to incorporate expectations of future wholesale electric prices and demand patterns (load growth, seasonality, etc.). That information would be important to determine the value of any additional generation from the hydro plant upgrade. On the other hand, it will not be known what the future streamflows will be and analysts usually rely on historical streamflow patterns to 'replay' the future with different historical streamflow patterns to get a sense of the range and variability of the value of the upgrade. Similarly, wind integration studies are often designed to capture some future state of the power system (loads, market prices, future generating resources) while replaying historically observed wind conditions.

Interactions between variations in load (demand for power) and wind generation are especially important in determining the cost of integrating wind. It is therefore important to select representations of both wind and load data that are appropriate to capturing the effects one intends to measure. The two most important effects revolve around the need for holding reserve generation sufficient to cover the net fluctuations of wind and load, and the need to determine the amount of peak demand wind generation can reliably serve. Capturing the important interplay between wind and load presents some special difficulties in representing the two.

While it is common practice for planning organizations to develop projections of load for future years, those projections do not usually incorporate sufficient information to create data sets of the hour-to-hour or within-hour variability. This likely leaves the analyst using historical load data, potentially scaled up to represent future load growth if it is necessary to determine costs for a future year. Relying directly on forecast load data for a future year runs the risk of failing to capture important statistical relationships such as its natural variability (hour to hour, minute to minute, day to day) with respect to the wind, and any potential correlation with wind generation—especially during load extreme events.

Using historical load data, potentially scaled up (or down) to simulate a future year necessitates using historical wind generation data for the corresponding historical year. Given that the wind resource under examination may not exist at the time of the analysis, much less some historical earlier period, this presents another obstacle. In the best of circumstances, historical wind speed data (from which generation might be derived per the last chapter) may be available for a recent historical year. In that case, the wind and load data can be taken over identical historical periods so that the variability of loads and wind will be matched. Matching wind and loads over the same time horizon is important for determining needed reserves, but is crucial for estimating the potential value of wind contribution to meeting peak demand. The reason for the latter is the effect described earlier, in which load extremes due to either abnormally warm or cool weather are often accompanied by large-scale high-pressure weather systems that bring calm winds.

Both the need for reserves and the computation of contribution to meeting peak demand are by nature statistical analyses. Given that, it is likely that in a single year, or even 2 or 3 years, there may be relatively little data to develop a precise evaluation of either the reserve levels or the peak demand contribution. In other words, it is likely that the resulting estimates will be relatively rough, with large uncertainties. Statistical methods can be used to quantify the uncertainty, but only large data sets can actually increase the precision of the results. It may be possible to use historical measurements and numerical weather models to reconstruct longer historical records on which to perform such analyses and reduce uncertainties. That approach may be especially useful in capacity contribution studies where wind behavior on the scale of an hour or more is important. The reliability of the output of such models on sub-hourly timescales may not be sufficient to use for studies of reserves. This is an area ripe for further investigation, however.

Easier to accomplish, but perhaps more difficult to interpret, is a strictly historical study of loads and wind over a more limited period (at least a year). The virtue of such a study is that the data are much more reliable and the results more understandable. That clarity of results may come somewhat at the cost of having to extrapolate the results into future years. Replaying historical loads and resources in the model using estimates of future wholesale electric and fuel prices is probably necessary to gain a picture of how the costs may evolve over time. This, however, makes it more of a hybrid analysis as described above.

Performing a purely forward-looking study, as is often done for planning purposes, is virtually impossible since the detailed temporal behavior of loads and wind on sub-hourly bases cannot be known. It might be possible to develop a complex model of wind and load behavior that could suffice. Such a model would probably be a stochastic function of variables such as hourly temperatures, market prices, behavior of specific load types (e.g. air-conditioning, irrigation, etc.), as well as larger-scale weather phenomena that affect both wind and loads. It would have to take into account the specific time dependencies and interdependencies of load and wind. This is another area where additional research could be fruitful. Efforts at creating stochastic wind speed or generation data have not generally succeeded in capturing the complex interactions among wind projects, appropriate statistical behavior at all timescales, and the important interrelationship with load and load variability.

5.3 REPRESENTING WIND AS LOAD REDUCTION OR FIXED GENERATION LEVELS

For the most basic economic analysis of the value of wind generation on a system, a study may be run that simply compares system costs without wind generation to the system costs after reducing the time series (e.g. hourly) of loads by the amount of wind generation. Such an analysis implicitly assumes (perhaps correctly) that the wind generators will not be actively controlled in any way. The usefulness and accuracy of such an analysis will depend on the sophistication of the load and wind time series used, the time step of the dispatch model (minutes, hours, days, weeks, or months), and the sophistication of the model logic for holding and dispatching reserves. Typically, time steps are an hour or more and reserves needed for the intra-hour variability of loads and wind are not explicitly modeled. It is usually necessary to estimate the intra-hour reserve requirements as a separate step outside running the model. This is a usual and acceptable course of action. Methods for estimating the reserve requirements are covered in the next chapter.

The nature of the wind generation time series determines the accuracy of this type of analysis. For example, if wind is represented as a single number per month representing the expected generation, the analysis will contain no information with respect to the cost of the variability of wind on smaller timescales. For example, the hourly variability of the wind may cause less efficient operation of thermal units than would happen under a fixed, known quantity for a day or month. Similarly, a power system may need to enter into additional hourly (or sub-hourly) market transactions with attendant transaction costs (bid–ask spreads, transmission, lost opportunities, etc.).

Some studies have attempted to isolate costs associated with the variability of wind by running multiple studies with different representations of the wind generation. Studies contrasting system costs with an hourly varying wind resource representation and system costs assuming a more constant energy source can highlight the cost of the variability itself—separate from the direct value of the energy produced and the cost of holding intra-hour reserves. This is discussed more fully in Chapter 8.

Representing wind as a load reduction or a fixed generation schedule (as a contract purchase or run-of-river hydro generation might be represented in a dispatch model) misses some of the potential value of wind resources. Most modern wind turbines can be controlled and are in fact controlled on systems where wind represents a significant fraction of the available generation. Moreover, it may be more economical to limit wind generation levels than to carry reserves sufficient to account for any conceivable, if infrequent, rapid change in output—at least increases in output. Ignoring the ability to control wind generation will not be a significant source of error for systems with relatively little wind, and might be reasonably overlooked. However, a fuller treatment of wind would represent wind's ability to limit output, albeit at some cost.

In no case should wind be allowed to inadvertently cause a thermal unit with long startup period to be displaced and risk resource inadequacy when the wind falls off. For example, wind coming up rapidly during light load hours should not be allowed to cause the model (or in actual operations) to de-commit a coal plant that might then be unavailable to serve load during the morning load ramp. For systems with relatively little wind, this is not a likely scenario. For systems with large amounts of wind, it may be necessary to limit the wind or take other measures to avoid this situation. In any case, the analyst should ensure that representing wind generation as a simple load reduction does not result in any adequacy issues. Similarly, it would be imprudent to allow wind to de-commit large thermal units in actual operations such that an adequacy problem could arise.

A potential weakness of the load reduction approach is that dispatch models will make unit-commitment decisions based on looking ahead in time at the expected market prices and load conditions. Representing wind as a load reduction may give the model unit-commitment logic too much certainty about the future. In other words, the model may look ahead at a situation where net demand is very low (as reduced by wind generation) and shut down a generator that may not have been shut down in actual operations due to the vagaries of wind generation—that is, forecasts that may not be sufficiently reliable to assure the timing of the wind generation.

A more complete and accurate representation would allow the displacement of wind at some incremental cost and would take account of the relative accuracy of wind generation forecasts at different points in time. Representing wind as a generation source is explored below. Perhaps not surprisingly, there are advantages and disadvantages. Representing wind as a load reduction has drawbacks—overly prescient thermal unit schedules and the inability to take account of the need to limit wind generation at times. Nevertheless, it is an important and useful tool, especially in cases where the overall wind on the system is relatively small.

5.4 REPRESENTING WIND AS AN EQUIVALENT THERMAL GENERATION STATION

A more complex representation of wind may be accomplished by treating wind generators as quasi-thermal units for modeling purposes. One advantage is to be able to make use of the existing model logic for limiting the output of a thermal generator at times. Such limits are most commonly exercised by model logic when the marginal cost of generating an additional megawatt-hour of energy is lower than the variable cost of running a particular generator. In other words, a particular generator will reduce its output when there are other opportunities to produce the energy from a less expensive source while still meeting the overall demand.

Of course, economic dispatch models generally assume that the fuel for a thermal generator is unlimited, and it may be difficult to capture the variability of wind generation strictly through adjusting model inputs. One way to do this might be to enter the maximum generating capability (a common model variable) as a time-varying pattern equal to the available wind generation. Similarly, a model might allow for time-varying maintenance schedules that may allow a de-rated capability to be entered. It will be important to run test studies to ensure that the computer model is respecting the generation limits correctly, and that the additional run-time burden is not unacceptably increased.

If run time becomes an issue, it may be addressed by combining the effects of multiple (potentially all) wind generators into a single generating unit. However, this will come at the expense of being able to evaluate the net effects of individual wind plant limits. For example, limiting every wind generator ramp rate to 10 MW/min is not the same as limiting the total output of all the wind projects to 10 MW/min.

It will be important to associate a variable cost to the wind generator. This is necessary to take advantage of economic dispatch models' ability to reduce generation at the wind generators when economically advantageous to do so. There are several reasons why the displacement of the wind resource may not be costless. For example, in cases where the wind is under contract from one party to another, failure to deliver generation may trigger payments of liquidated damages to the purchaser. Even without such provisions, lost value may accrue through a reduction in the production of renewable energy credits that may count toward government-mandated renewable energy standards, or through the loss of tax credits designed to provide incentives for wind generation. Whether or not these values (lost income, damages, tax credits, REC value) should be accrued in the model economic analysis is a decision for the analyst that needs to be based on whose perspective the model costs are meant to reflect.

Representing wind projects as thermal generating stations having time-varying maximum output capability allows for capturing the economic displacement of wind projects, but unfortunately may still suffer from the problem of the model having too much foreknowledge about the amount of wind available to it over future hours, days, or even months. As already mentioned, too much certainty about future wind generation can cause the model to take steps to shut down other generating units for economic reasons, which actual operators might not do because of the real-life uncertainty in wind generation. As wind generation forecasts improve, this may become less of a drawback. Given the current state of wind forecasting, the foresight issue is significant enough to be addressed if possible. It may be necessary to run models twice—once assuming a forecast level of wind (with attendant forecast error) and a second time with the modeled wind. This technique is discussed further in Chapter 8.

Models tend to have rather extensive logic to model the complexities of thermal generation stations' (power plants powered by natural gas, diesel, coal, or nuclear fuels) outage states. That logic is attractive to analysts desiring to model wind generation using logic designed for thermal generation stations. In this scheme, the variability from one time period to the next (e.g. hours or minutes) might be modeled as availability state transition probabilities. Each availability state represents an output level for the wind project. Transition probabilities represent the probability of moving from a current generation level (availability state) to another. Some dispatch models allow the definition of multiple partial outage states to represent power plant de-rates (partial outage level) arising from equipment failures. Such models allow users to define the de-rated states with transition probabilities between the states. The transition probabilities generally represent the probability of entering a new failure state, or else repair of an existing failure. Table 5.1 shows an example of generation states and transition levels between 15-minute periods, and Figure 5.1 illustrates the same data graphically.

As an example in interpreting the meaning of Table 5.1, assume that at some particular 15-minute time period the generation at the project is at 15% of nameplate capacity. In that case, the 'initial state' would be represented by the second row of data, labeled '10–20%' in the first column on the left. The probability of the generation dropping to between zero and 10% in the next 15-minute period would be 21.63%, as depicted in the '0–10%' column. In other words, when this project is generating at between 10% and 20% of nameplate, there is a 21.63% chance it will drop in the next 15-minute period, a 58.63% chance of remaining roughly at

TABLE 5.1 Transition Matrix (Probabilities)

	End State									
Initial State	0–10%	10–20%	20–30%	30–40%	40–50%	50–60%	60–70%	70–80%	80–89%	90–100%
0–10%	94.13%	4.66%	0.76%	0.21%	0.08%	0.05%	0.06%	0.01%	0.02%	0.02%
10–20%	21.63%	58.63%	14.54%	3.09%	1.11%	0.46%	0.16%	0.23%	0.07%	0.10%
20–30%	1.94%	24.80%	48.87%	16.68%	5.05%	1.62%	0.41%	0.41%	0.23%	0.00%
30–40%	0.79%	5.87%	25.23%	40.63%	17.72%	6.09%	2.20%	0.96%	0.45%	0.06%
40–50%	0.44%	1.33%	6.53%	25.35%	36.82%	19.96%	6.53%	1.84%	1.08%	0.13%
50–60%	0.19%	0.39%	1.75%	7.19%	23.53%	35.00%	22.03%	6.74%	2.72%	0.45%
60–70%	0.00%	0.30%	0.48%	1.98%	7.25%	21.57%	37.51%	21.63%	7.37%	1.92%
70–80%	0.05%	0.11%	0.27%	0.55%	1.54%	6.32%	21.87%	40.16%	24.07%	5.05%
80–90%	0.00%	0.03%	0.06%	0.15%	0.45%	1.27%	3.50%	14.82%	57.40%	22.33%
90–100%	0.02%	0.00%	0.00%	0.04%	0.02%	0.13%	0.45%	1.24%	14.00%	84.11%

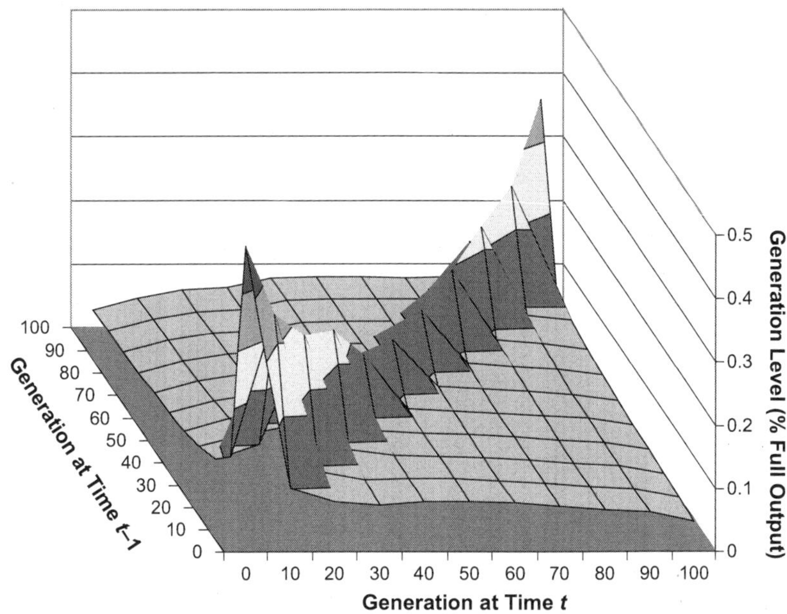

FIGURE 5.1 Generation state transition probabilities.

the same generation (in the 10–20% column), and only a 0.10% chance of increasing to within 10% of full output (90–100% column).

Using transition matrices (more broadly termed Markov chain processes) is very tempting to the analyst as it embodies the statistical nature of wind output behavior from one time period to the next. Unfortunately, this technique suffers from some potentially fatal shortcomings. It should be noted that the example in Table 5.1 and Figure 5.1 represents data for a single project for an entire year (15-minute periods). While broadly representative of the behavior over a year, it would only be reasonably usable if the probabilities do not change over the hours of a day, or the seasons of the year. Most wind projects do show seasonal and diurnal variations that are important to capture. These could be captured by imposing a number of transition matrices, each one representative of a particular hour of the day in a particular season of the year. This complicates incorporating the technique into commercially available dispatch models that may not have the ability to nominate multiple transition matrices.

There are other potential problems that are more severe. While a transition matrix is useful for modeling a single project, it is significantly more complex when multiple wind projects are to be modeled with significant correlations among them. The matrices quickly become multi-dimensional, where the transition probabilities are conditioned on not just the current output of the project, but also the current output of all the projects.

Finally, to get a reasonable granularity in output levels (e.g. to the nearest megawatt), the multiple (by hours of the day and seasons of the year) multi-dimensional (to account for the state of other generators) matrices rather large, and the resulting model run times too slow to be of practical use with current generation computers and software. It is possible that this limitation will eventually relax as computers continue to get bigger and faster.

5.5 SUMMARY

Economic dispatch models commonly available to analysts do not generally provide the ability to:

1. Generate data to represent temporal behavior of wind generation.
2. Represent dispatch characteristics of wind generation.
3. Incorporate the importance of accuracy of wind and load forecasts versus schedules.

As a result, the analyst is often left having to approximate the behavior of wind generation in the models. These approximation methods include:

1. Treating wind generation as a load reduction, contract sale, or run-of-river hydro units.
2. Treating wind as a thermal generator with either:
   a. A time-variable maximum output capability or de-rate schedule that is known and fixed at the beginning of the model runs.
   b. A stochastic variable with generation levels varying in a pattern to simulate the uncertainty of wind generation experienced in actual operations.

Each of these techniques has strengths and weaknesses. Treating wind as a load reduction is the simplest to implement and to validate, but does not reflect economic displacement of wind. At low levels of wind penetration, there is little likelihood of economic displacement, and this approach is a reasonably good approximation. A potentially significant weakness of this method may be the model's ability to see into the future too well and displace thermal units at times when it would not be done in actual operations due to the uncertainty in wind forecasts.

Representing wind as an equivalent generating resource potentially provides a more realistic treatment of wind displacement and can be a useful approach for systems with larger amounts of wind. Available generation can be established as a time-dependent pattern outside the dispatch model. Depending on the particular dispatch model's unit-commitment logic, the model may still assume an unrealistic foreknowledge of generation levels.

A further enhancement would be to represent the generator availability as a stochastic variable. Such an enhancement comes at the expense of model run time, and a potentially data- and analysis-intensive effort to develop an accurate representation of the statistical properties of the wind generators. Chapter 8 explores an alternative to this approach in which two sets of wind generator availability data are prepared: one representing the expectation of wind generation and the other the wind generation experienced. The idea here is to have the dispatch model make unit-commitment decisions based on a forecast of wind generation, fix the unit commitment in that run, and then re-run the study with the actual wind data.

It is unfortunate that such machinations are necessary due to the current inability of dispatch models to adequately address the unique characteristics of wind generators. It is of course hoped that model vendors will quickly respond to this deficit and incorporate necessary logic changes that would allow a more straightforward approach to modeling wind on power systems.

Chapter six

Power System Incremental Reserve Requirements

He is the best sailor who can steer within fewest points of the wind, and exact a motive power out of the greatest obstacles.

Henry David Thoreau, 1817–1862

Assessing the cost of accommodating wind on power systems is primarily driven by the cost of providing additional generating reserves required to meet the uncertainty and variability of wind generation. This chapter addresses how to determine the extent of the incremental reserve requirement and lightly touches on valuing the opportunity costs associated with holding reserve generation. At the most general level, reserve refers to generating capability purposely arranged in excess of the expected need for generation to accommodate unexpected behaviors of demand and generation. Power systems 'schedule' generation based on the expectation (forecast) of demand and generation. Each generator in the system is assigned to generate at its schedule over the operating period. To the extent that the actual loads and generation differ from the schedules, reserve generation is available to make up the difference.

6.1 PRINCIPLES OF RESERVE REQUIREMENT ANALYSIS

Before launching into how incremental reserve requirements for wind generation are determined, it will be important to keep in mind three fundamental points:

1. The purpose of holding additional reserve generation is to ensure the adequacy and reliability of the power system.

Valuing Wind Generation on Integrated Power Systems. DOI: 10.1016/B978-0-8155-2047-4.10006-7

2. Reserve requirements stem from both variability and uncertainty of wind generation—both characteristics are important, but separate, factors to be considered.
3. Incremental reserve requirements due to wind generation cannot be determined separately from consideration of reserve requirements for loads.

Exactly how these guidelines are incorporated into the analysis will become clear in the description of how the analysis is undertaken. First, it will be worth examining what these points mean and why they are important cornerstones to any reserve requirement analysis.

6.1.1 Incremental reserves to ensure reliability

Prudent utility operations require some generating capability held on unloaded units (i.e. generators not operating at maximum output capability) to respond to unpredicted needs that may arise within an operating period. Such needs include sudden and unexpected increases in system demand, or power system component outages (also called 'contingencies'). With the advent of wind generation, one such use for reserve capability is to compensate for unforecasted or rapid decreases in wind generation within the operating period.

It is not prudent, economic, or even possible to design a power system with sufficient reserves to respond to any conceivable set of circumstances that may require additional generation. Reliability standards are invoked to provide guidance in determining a prudent level of reserves to be held. For example, the North American Energy Reliability Corporation (NERC) has established Control Performance Standard 2 (CPS 2) requiring that 10-minute average area control errors[1] remain within established limits (called 'L10 bands'), 90% of the time in each month.

The purpose of holding reserves is to remain in compliance with all such reliability standards. It is typically up to power system operators to determine the specific amounts and types of reserves necessary to remain in compliance as each utility can have different needs depending on the specific characteristics of their loads and resources. Methods already in place to determine reserve requirements for existing loads and resources may be applicable to determining incremental requirements due to wind. However, such methods are often the product of years of utility personnel experience with existing types of loads and generators, and may not be applicable to systems with large or growing amounts of wind generation. Any incremental reserve requirement that may be deemed necessary to accommodate wind variability and uncertainty should be determined on the basis of meeting power systems reliability

[1] Area control error measures the extent to which scheduled loads and resources differ from the observed levels. It is discussed more fully below and in the Glossary.

targets. The need to meet reliability targets provides the basis for any reserve analysis.

Ideally, the analysis leading to an incremental reserve determination for the system with wind would not be different than the specific utility's standard practices for making such determinations without the wind—to accommodate the variability and uncertainty around system demand and other generators. As will be discussed more fully below, these kinds of determinations can be complex, and utility operators, especially in smaller systems, may employ rules of thumb that keep the system reliable without complex analyses. Generally, applying the traditional reserve margin rule (e.g. providing nameplate capacity reserve margins as a fixed percentage of expected peak load) will not work when applied to a system undergoing significant changes to its basic characteristics due to the addition of wind generation. It is critically important to employ a methodology for determining reserves that is directly related to maintaining reliability standards. Doing otherwise risks vastly over- or underestimating the reserve requirements for wind.

6.1.2 Distinct importance of variability and uncertainty

Variability and uncertainty are not the same thing and a complete analysis of wind integration costs will capture both the costs deriving from the variability of the wind and the uncertainty with which it is forecast. Although this is often intuitively clear to analysts, the two are sometimes conflated. To illustrate the difference, consider the case where a utility sells its wind generation, to the extent possible, into a liquid market that transacts over hourly operating periods. Figure 6.1 illustrates the needed reserve unit operation to balance the system over 8 hours in the case where the wind generation levels are known ahead of time with complete certainty and the scheduled delivery equals the average generation over each hourly operating period.

In this particular example, 95% of the within-hour variability lies between −17.4 and 22.8 MW (negative values represent reducing generation that would otherwise be used to meet load). Conversely, consider the same case where there is a significant forecast error. Figure 6.2 illustrates the same wind data where the delivery schedules are based on wind output forecasts with a forecast standard error of 30%.

Although the data here are anecdotal, it illustrates that the need to maintain and operate reserve generation increases with forecast error. In this case, the need for reserve units increases 63% for the incremental dispatch (from 22.8 to 37.2 MW) and 66% for the decremental dispatch (from −17.4 to −28.9 MW).

The distinction between variability and uncertainty is more than semantic. Levels of reserve generation needed to accommodate variability are primarily affected by the wind resource variability, but are also affected by the length of the operating period. For example, the variability

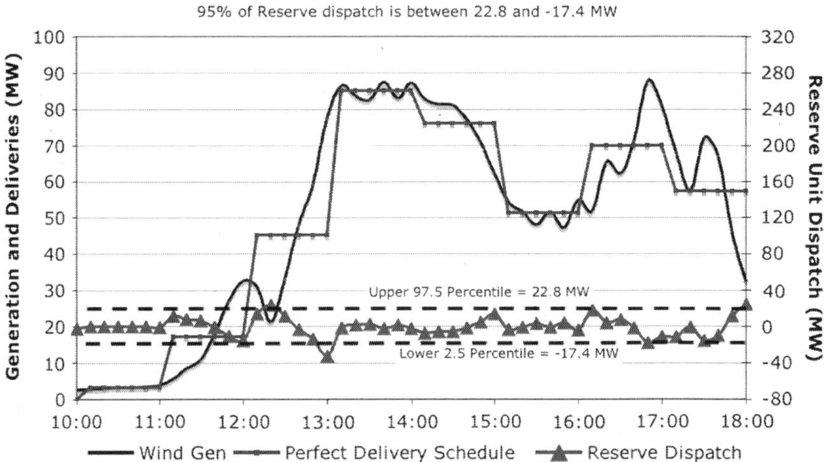

FIGURE 6.1 Reserve unit dispatch where schedules are based on perfect foreknowledge of wind generation levels over hour-long operating periods. Reserve units must increase or decrease output to compensate for the within-hour variability of wind generation.

experienced over a 10-minute operating period is considerably less than that over a 1-hour period. Conversely, reserve requirements associated with forecasting accuracy are very sensitive to forecasting technology, processes to rapidly convert forecasts to schedules, and market rules for the timing of submitting schedules. Variability and uncertainty separately and differently contribute to the cost of accommodating wind generation.

This example is illustrative of the relative importance of variability and forecast accuracy on reserve requirements—they are roughly equally important depending on specific circumstances (length of operating period, market liquidity, variability of the wind resource, and accuracy of existing forecasting services). It is crucially important to recognize and fully take account of the separate effects of variability and forecast accuracy.

6.1.3 Reserve requirements depend on both load and wind characteristics

Reserves are needed to meet the variability and uncertainty of power demand, regardless of the state of wind generation on the power system, or indeed regardless of whether there is any wind generation on the power system. Adding significant amounts of wind increases the need for holding reserves, but the interaction between load and wind reserves is important. In general, the two reserve needs cannot be independently determined and added together to find the total reserve requirement. This is a case where the total is usually significantly less than the sum of the parts.

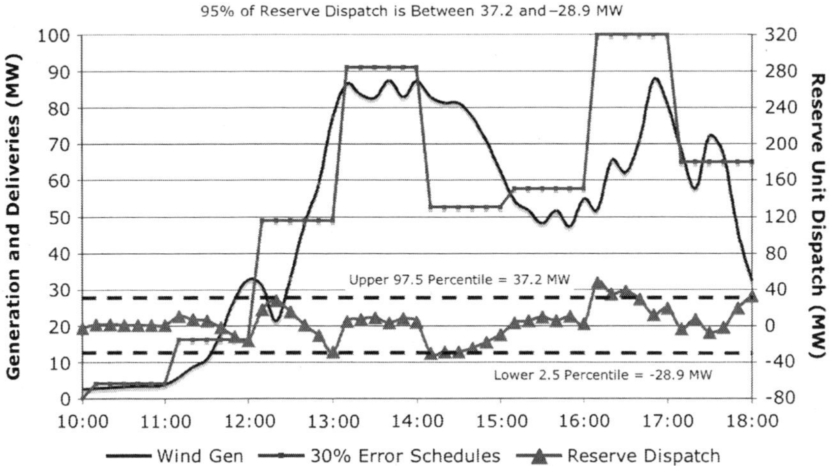

FIGURE 6.2 Reserve requirements increase when wind forecast error is included.

As an example, Figure 6.3 shows the output of reserve generators (labeled 'Schedule Error') that would be required for sample loads on a particular day. Schedule error is defined as the difference between the scheduled generation (based on forecasted load over the operating period) and the observed load. Note that 90% of the need for reserves lies between −176 and 180 MW in this sample. Negative values of schedule error are achieved by reducing the output of operating reserve units below their scheduled levels by the negative value.

For this same period, a similar analysis could be done separately for wind generation and wind generation schedules. The implicit assumption here is that all of the difference between the wind schedule and the actual wind generation would be met by reserve units dedicated to that purpose. Such an analysis is shown in Figure 6.4, with wind generation data from the same time period as the data in Figure 6.3. Note that 90% of the need for reserves to balance wind in isolation from load is between −158 and 97 MW in this sample.

If the needs for load and wind reserve dispatch were additive, one would expect that the total reserve requirements at the 90% level would be −334 MW (−158 + −176 = −334 MW) and 277 MW (180 + 97 =277 MW). In fact this is not the case. Figure 6.5 illustrates the tendency for load and wind schedule errors to cancel one another out much of the time, and that the likelihood of errors to be both near maximum levels and in the same direction is relatively low. As a result, the total reserve requirement is significantly lower than would be expected from separately calculating the requirements for loads and wind and then summing the two. At the 90% confidence level, the total reserve requirement is between −244 and 207 MW in the sample data—values that are a quarter

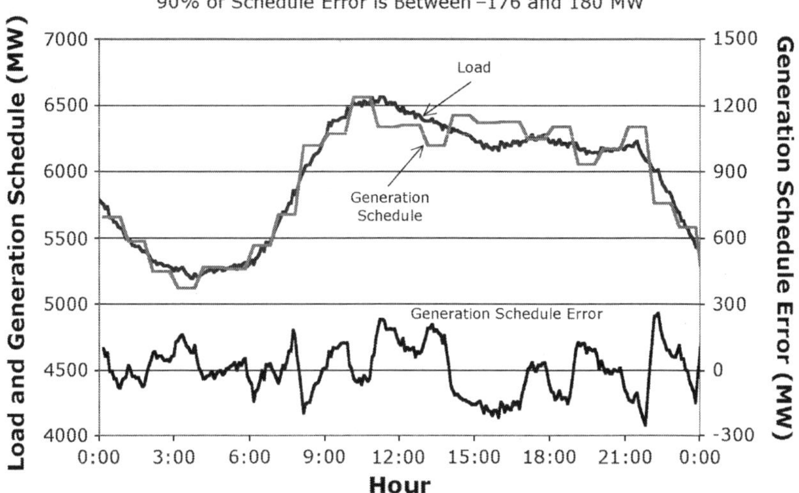

FIGURE 6.3 Load and matching hourly generation schedule, with the resulting schedule error shown on the secondary axis. Reserve generation units would need to be dispatched to make up for the schedule error.

to a third less than simply adding the separate reserve requirements in this example.

Although this was done with sample data, it is a general result. The only way in which the reserve requirements could be additive is if there is a high, nearly 100%, correlation between the load and wind schedule errors. This is unlikely to be the case—even if load and wind generation are significantly correlated, it is unlikely that the schedule errors would be correlated.[2] It is therefore crucially important to calculate the incremental wind reserve requirement together with an analysis of load reserves.

An interesting, and potentially useful relationship illustrated with this sample data is that the 90% schedule error values in Figure 6.5 may be closely approximated from the independent values in Figures 6.3 and 6.4. For example, if the upper bound values (180 and 97 MW) are squared and then summed, the square root of that result is close to the upper bound found in Figure 6.5:

$$\sqrt{180^2 + 97^2} = 204 \text{ MW} \approx 207 \text{ MW}$$

Similarly, for the lower bound in Figure 6.5:

$$-\sqrt{176^2 + 158^2} = -237 \text{ MW} \approx -244 \text{ MW}$$

[2] It is possible but rare that a large-scale weather front causes both load and wind generation scheduling errors that are additive. Such events would not likely dominate the statistical behavior of most systems.

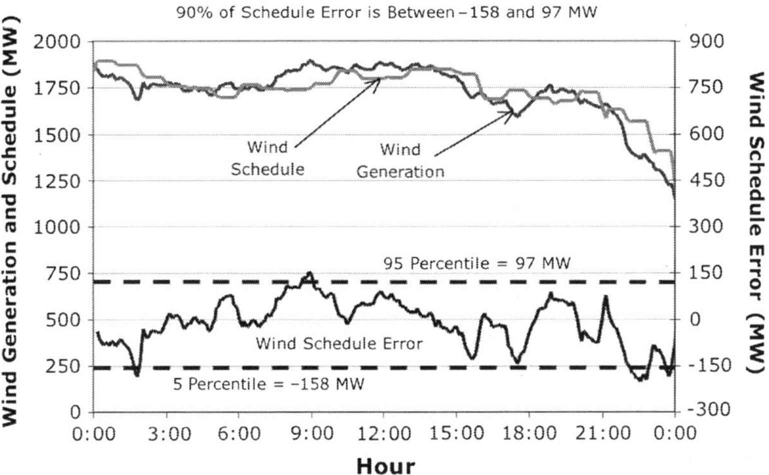

FIGURE 6.4 Wind generation and wind generation schedules with the wind schedule error taken as the difference between the two. Reserve units would be dispatched to meet the wind schedule error without other generation and loads on the system.

Although approximate, this provides a useful shorthand for estimating the joint effects of wind and load separately. Why it works is explored later in this chapter.

MATCHING GENERATION AND DEMAND

Reserves are used by power systems to maintain system reliability by adjusting generator output to match demand. It is sometimes said that power system operators use reserve generators to ensure that the amount of energy generated at every moment equals the amount consumed. While this model is useful in conjuring the general idea, the reality is somewhat different. The physical law of energy conservation maintains that the amount of energy produced is always equal to the amount consumed. Power consuming and generating devices are designed to work within a range of voltages and frequencies. In North America, the range is typically around 120 volts and 60 hertz (240 volts and 50 hertz in Europe and many other countries). When demand for power is different from the supply of power, these system parameters vary from their design values.

Take, for example, a 20-watt incandescent bulb that is designed to consume 20 watts when connected to a power source operating at 120 volts and 60 hertz. If a generator capable of producing only 15 watts is connected to the nominally 20-watt light bulb, it will not be able to maintain the nominal 120 volts and 60 hertz—the voltage will drop and so may the frequency until the light bulb consumes no more than the 15-watt generator capability. It is possible for the generator or load (light bulb in this example) to fail due to the under-voltage and frequency power. However, if the generator is operating differently from, but reasonably near, 120 volts and 60 hertz, the bulb will simply consume a slightly

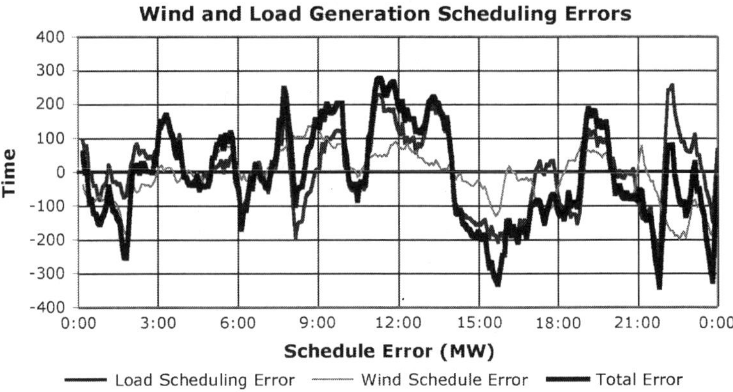

FIGURE 6.5 The effects of netting load and wind schedule errors.

different amount of power than the designed 20 watts, resulting in a somewhat brighter or dimmer level of light output.

Similarly with a complex power system, if the demand (not necessarily the consumption) is greater than the capability of the operating generators, the frequency of the power will generally fall below the nominal design levels. This is an indication to the power system operators that additional generation (or load dropping) must occur to maintain the system within acceptable tolerances. Conversely, if the frequency of the power system increases, generation must be reduced (or loads added) to maintain the designed frequency level.

Although it is not correct to believe that power system operators use reserves to make critical adjustments to match generation and consumption at every moment, it is important to note that there can be extreme consequences to operating the power system outside design parameters. Electric motors and generators operating outside the acceptable ranges may heat up internally and fail at lower or higher frequencies and voltages. Many devices automatically disconnect from the power grid at inappropriate frequencies or voltages to protect the equipment from terminal failure. It is common for electric power system components to be so equipped. When power system parameters fall outside the acceptable ranges, the automatic cutout of generators can result in a cascade of power plant shutdowns that can take down the entire power grid. Power system operators avoid such generalized blackout scenarios at all costs.

6.2 RESERVE NOMENCLATURE

Power systems operate under a great deal of uncertainty and variability in generation and consumption. Demand for power continually changes over timescales of seconds to minutes, hours, days of the week, and seasons of the year. In addition, generators and transmission lines are subject to sudden and unexpected failures that may necessitate rapid and significant intervention to avoid widespread power outages. Accommodating all of this variability and uncertainty necessitates maintaining reserve generators that can respond as needed. Reserve generators have the characteristic that they have at least some potential to increase, and in some cases decrease, generation levels as needed. Terminology is not consistent with respect to different categories of reserves. For the purposes of this book, reserves will be divided into two broad categories: 'planning reserves' and 'operating reserves'.

6.2.1 Planning reserves

One example of maintaining reserves relates to meeting the peak demand periods which generally coincide with temperature extremes that drive additional demand—typically from air-conditioning loads in summer months or electric heating loads in the winter months. Whether summer or winter peaking, utilities look prospectively at the next season and estimate the level of the peak demand. Operators will arrange for generating facilities to meet the expected demand for the peak season. Both because the peak demand could be higher than expected (due to temperature extremes or higher than expected growth in the customer base) and power generators may fail, it is standard practice for operators to arrange for generation above and beyond the expected demand levels to provide reliable service in the case of generator failures and/or higher than expected demand. The generators made available above and beyond the expected need are referred to as reserve generation, and the level of the deemed additional need is the 'planning reserve requirement'. Wind generation is generally expected to have relatively little effect on the need for planning reserves. A fuller examination of wind's contribution to meeting peak demand and planning reserves is presented in Chapter 10.

6.2.2 Operating reserves

Terminology relating to reserves varies widely within the USA,[3] and is generally different between the USA and Europe as well. For the purposes of this book, 'operating reserve' will refer to a broad category of reserves, but other usages of the term exist that use it to mean a specific need for reserves—those necessary to replace a sudden loss of a major power system component. Such reserves will be referred to as 'contingency reserves' here.

---

[3] The differences in the USA may stem from different definitions used by two important organizations—the usage adopted here is meant to be consistent with the North American Energy Reliability Corporation (NERC) 2002 Policy 1. However, the Federal Energy Regulatory Commission (FERC) Order 888 defines operating reserve as: 'Extra generation available to serve load in case there is an unplanned event such as loss of generation.' This latter usage is termed 'contingency reserve' in this book.

For the purposes of this book, 'operating reserve' will be used to represent the broadest category of reserve generation needed to keep the power system within acceptable reliability criteria on a real-time basis throughout each day. This includes capability above firm system demand required for regulation, forecasting error, and equipment-forced and scheduled outages. Other terms will refer to subcategories of the purposes for holding operating reserve, and subcategories of the characteristics of generators that supply operating reserve. These more specific terms includethe following:

Reserve terms relating to purpose

- **Contingency reserve** represents the need to hold reserve in the case of a significant power system component outage such as a generator or transmission line. In the USA, at least half the contingency reserve must come from spinning reserve generators.
- **Regulation reserve** refers to reserves needed to maintain power system balance due to the fluctuations of load and wind generation that occur in a timeframe of a few seconds to a few minutes. The net energy transacted over an hour to provide regulation reserve is generally close to zero because balancing involves both increasing and decreasing generation levels multiple times throughout an hour.
- **Load following (or just 'following') reserve** refers to the need for generation to increase and decrease over a period of many minutes or hours to follow the diurnal pattern of demand and slower movements of wind generation. Load following reserve usually entails a net energy transfer (positive or negative) over a period of an hour, but is expected to net to zero over periods of a day or a month.
- **Imbalance reserve** refers to generation that must respond to the average difference between scheduled generation (or load) and actual generation (load) over the operating period. Imbalance reserves entail net energy transactions over periods of hours or even days, but would be expected to net close to zero over periods of weeks or months.

Reserve terms relating to generator characteristics

- **Primary**, **regulating**, or **spinning reserves** are generators capable of responding (increasing or decreasing generation levels) within one or a few seconds. Such units must be energized and synchronized to the power grid.
- **Secondary** or **non-spinning reserves** are generators capable of responding within 10–15 minutes.
- **Tertiary** or **supplemental reserves** are generators capable of responding on longer timescales of typically 30 minutes or more, usually to reduce the burden on other reserve units after a significant contingency event.

Reserves are held principally to maintain power system frequency within an acceptable range. Voltage may also be affected by an under- or over-abundance of power generation, but is generally managed separately from increases or decreases in energy production or consumption. Generators usually provide the reserves needed to balance power systems and maintain reliability criteria. However, the same services could equally be performed (and potentially at lower cost) by energy-consuming equipment—i.e. loads. Loads such as electric heaters and lights may have a broad range of energy inputs at which they may still perform the intended functions acceptably. Expanding the use of energy-consuming equipment to provide reserve services is explored more fully in Chapter 12.

In the absence of wind on a power system, the primary need for holding reserves arises from the variability and unpredictability of load, and for sudden power system equipment failures (contingencies). It is possible for reliability criteria such as frequency limits to be maintained even in the event of relatively extreme imbalances between a particular system's demand and generation of electric power. This can occur at the expense of service to consuming equipment (i.e. load is different from demand), energy stored in the power system itself (e.g. kinetic energy of rotating equipment) and neighboring systems that may be unexpectedly absorbing or providing generation. To minimize one system from inadvertently (or purposely) relying on its neighbors, regulatory bodies may impose limits on these unintended power transactions. A measure of the extent to which a system is out of balance is called the area control error (ACE), referred to previously. Regulatory bodies may impose limits on the allowable size of the ACE, in addition to, or separate from frequency requirements. Reserves are held and deployed as necessary to maintain such reliability standards.

6.3 DETERMINING NON-CONTINGENCY OPERATING RESERVE REQUIREMENTS

Determining reserve requirements is a process including the following steps:

1. Developing both wind generation and load time series potentially from either modeled or observed data—usually synchronized over identical historical time periods.
2. Establishing load and wind schedules (e.g. from forecasts) to be used over the power system's operating periods (e.g. each hour, 10 minutes, etc.), respecting the required notice period to submit the schedules.
3. Further dividing the remaining variability into relevant time periods (a few seconds, minutes, or tens of minutes) to determine the need for different types of reserves.
4. Determining the level of reserves necessary to maintain reliability criteria within acceptable bounds.

Developing wind generation time series over fixed historical time periods is discussed in Chapter 4. Load time series should be accessed from historical data capturing the within-hour variability for which reserves are needed. It is also important to include the effects of scheduled[4] wind generation and load. Reserves will be needed to capture not only the variability within the operating period, but also make up the average difference between the expected loads and the scheduled generation over the operating period.

Power systems routinely produce load forecasts that are used to set generation levels ('set points') based on submitted schedules. It may be possible to find forecasts made for a historical time period, or it may be necessary to reproduce schedules by recreating a forecast made up of the actual load level plus a random load forecast error that has the appropriate statistical characteristics (i.e. reproduces historical load forecast errors). Load forecasts out an hour or two into the future may be reasonably simulated by using a simple curve fit over the past few hours. Another method looks for a 'similar day' in historical load patterns from similar seasons.

Techniques for developing wind forecasts for a few hours into the future are different from load. Whereas it would be unusual for the load pattern on one day to be unrelated to that of the day before, this is not necessarily the case for wind. Wind generation patterns are not typically amenable to time series treatment such as curve fitting or similar day analysis, which relies on repeated patterns. It is not uncommon for a calm day to be followed by a day of relatively high wind energy output. Although some wind projects have diurnal patterns, or tend to produce more energy in some seasons relative to others, the patterns are irregular and not predictive for any specific day.

It is common to assume simple persistence forecasts for time periods of up to about 2 hours prior to the operating period for the purposes of determining reserve requirements. Persistence forecasts take a current level of wind generation and assume that the wind generation continues at that level through the operating period. For example, a forecast produced 1 hour prior to the operating period would simply be the wind output observed at that time. Although wind output can change rapidly over periods of tens of minutes to hours, for relatively short time periods, persistence forecasts may be sufficient. If, for example, the output is near zero at some time, the likelihood that it remains at that level, certainly over the next few minutes to an hour, is very high. Figure 6.6 illustrates the

[4] Scheduled load and generation may be thought of as forecasts of the respective quantities. One important difference is that while forecasts may be made at any time prior to the operating period (e.g. 1 week, 1 day, 1 hour, 1 minute), schedules represent the statement provided by the scheduler to the power system entity and are due at pre-selected times prior to the operating period. For example, schedules may be due 30 minutes prior to the start of each operating period. More accurate wind or load forecasts may become available after that time, but cannot be used to affect operations.

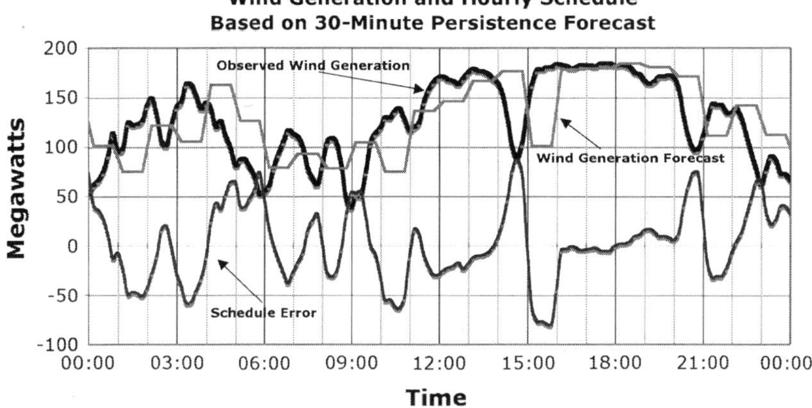

FIGURE 6.6 Hourly wind schedule based on 30-minute persistence wind forecast.

accuracy of a persistence forecast made 30 minutes prior to an operating hour.

Forecasts of wind generation and demand are used to produce schedules of wind generation and load. Those schedules are used to establish generator levels ('set points') and market transactions designed to balance the system as well as possible over the operating period. The difference between the generator set points and the actual load net of wind generation (and other less controllable resources) must be met by reserves.

For example, the load for an upcoming operating period might be forecast to be 2000 MW, and the wind schedule suggests wind generation at 75 MW. The balance of the power system will be set up to meet the load net of scheduled wind generation of 1925 MW. Generators will be sent signals to operate, in total, at that level. To the extent that wind and load vary from their respective marks, reserve generation must be available to balance the system to cover the differences, or 'schedule errors'. Reserve requirements are determined by finding a level that covers the errors at a selected reliability level. For example, if reserves must be available to encompass the errors 95% of the time, then the appropriate reserve levels are those that encompass 95% of the error terms.[5]

Such a distribution of theses errors is illustrated in Figure 6.7, where 6 months of 5-minute data were used to develop differences between the hourly scheduled wind and load levels and the observed values using 1-hour operating periods. The data are from the Bonneville Power

[5] More accurately, an adjustment may need to be made for the finite sample size of the selected historical period. In other words, the level of error terms at the 95th percentile level represents an approximation of the error of an unknown larger population. This is especially important where the historical period used represents just a few months, or the data are segmented as is described later in this chapter.

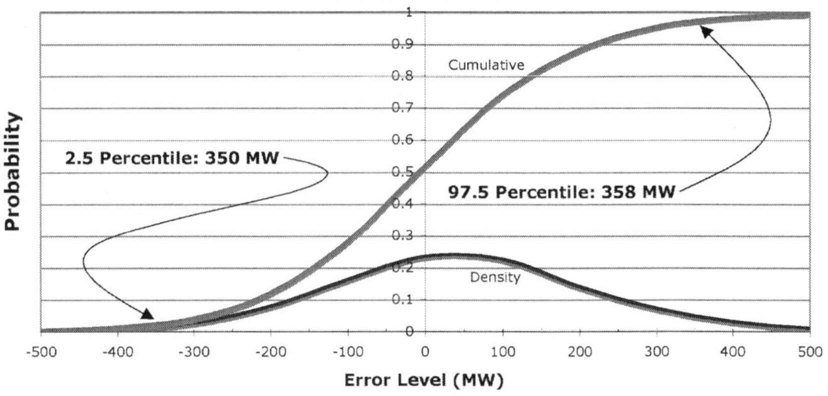

FIGURE 6.7 The distribution of the differences between the hourly schedules and observed wind net of load for a 6-month period on the Bonneville Power Administration system. Distribution has a mean of −4 MW and a standard deviation of 177 MW.

Administration with peak load of 10,762 MW representing July through December 2008, and peak wind generation of 1510 MW in that same period. The wind schedules were taken directly from the observed values 30 minutes prior to the start of each clock hour. A proxy of load schedules was produced by introducing an error term around each hour's observed load using a normal distribution with a standard deviation of 2% of the observed value. Schedules were ramped from one hour to the next beginning 10 minutes prior to the start of the operating hour through 10 minutes after the end of the operating hour.

As Figure 6.7 shows, 95% of the 5-minute errors fall between −350 and 358 MW, indicating the need for this system to maintain 358 MW of incremental generating capability and 350 MW of decremental generating capability. Decremental capability is the ability to reduce generation from the set-point levels on loaded units. Some generators may not have the ability to reduce generation in the appropriate timeframe, potentially because they do not possess the communication and control equipment necessary, or they may already be at their minimum acceptable generating level, or they may not be able to change output as rapidly as may be required. Although it is common to think of reserve requirement as an ability to relatively rapidly increase overall power generation, it is equally necessary to be able to reduce generating capability as well.

It is important to note that the results of this analysis represent the total amount of reserve needed for the combined effects of wind and load variability and uncertainty. The total (non-contingency) reserve requirements can be further divided by type of generating reserves (e.g.

regulation or load following), or to allocate reserve requirements to wind and load. Any segmentation of reserve requirements, either by cause (wind or load) or by type of reserve, must carefully consider how the subcategories sum to the total. For example, looking separately at the differences between hourly load schedules and observed load in the Figure 6.7 data, the 97.5 percentile level is 274 MW. The corresponding wind schedule error at the same percentile is 230 MW. It is clear that the total reserve requirement of 358 MW found in Figure 6.7 is considerably less than the sum of the corresponding wind and load percentile levels. This is a general result, and techniques for allocating reserves by cause and type are discussed below.

6.3.1 Segmenting reserve requirements by type

Reserves can be broken down into different categories representing the rapidity with which they can respond, or else their root cause (e.g. variability versus uncertainty). Regulating reserve is the most responsive, representing loaded units with control and communication equipment allowing automated response within a few seconds. Not surprisingly, this service is also the most expensive to provide and it is important to minimize the amount of reserve assigned to these units. Much of the movement of wind and load occurs over longer timeframes of 10 minutes to several hours. As a consequence, analysts may want to segment that part of the reserve that must be met by automated and rapidly responsive generating units, from that part of the reserve that might be met by reserves that respond on 10- or 30-minute bases.

Several methods of ferreting out the different reserve requirements are in use. One method is to smooth the errors over a timeframe of, say, 10 minutes assuming that the choice of regulating reserve level is sufficient to produce such smoothing. It would be necessary to start with data on as fine a timescale as regulating reserve, no greater than about 1 minute. Wind generation data in 1-minute time increments may be difficult to acquire, and may be difficult to deal with due to its sheer size—half a million data points for each year. As will be illustrated in the examples below, the minute-to-minute variability of wind is relatively modest. As a result, it is reasonable to separately estimate the effects on that timescale, while relying on 5- or 10-minute data for the remainder of the reserve analysis. For completeness, the analysis below assumes data at 1-minute increments.

To illustrate, one sample from a month of 1-minute wind and load data shows observed wind and load deviations from hour-long schedules range from −410 to +421 MW 95% of the time. Smoothing the 10-minute time blocks is found to be accomplished 95% of the time by holding fast-acting (regulating) reserves ranging from −72 to 70 MW—roughly 17% of the total non-contingency reserve requirement. Figure 6.8 illustrates the distributions of these values. While it would suffice to hold regulating

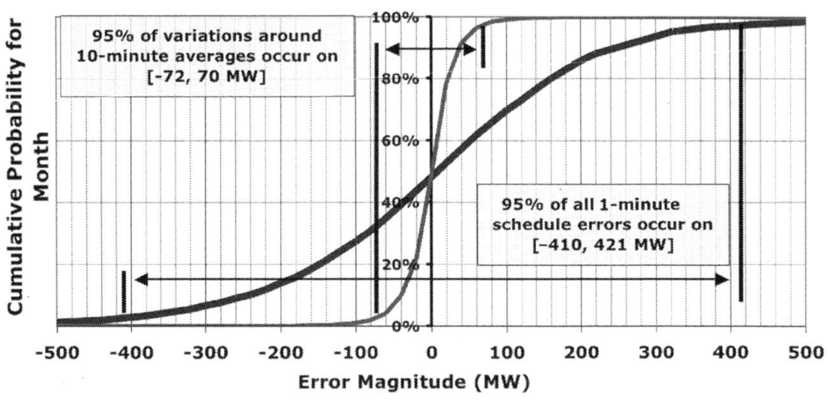

FIGURE 6.8 The distribution of the need for faster-acting reserves able to respond to minute-to-minute variations about 10-minute averages, versus the overall need for reserves to address variations from schedules. One month of data with a peak load of 7700 MW and maximum wind generation of 2100 MW.

reserves for the entire roughly ±420 MW, it may be more economically efficient to break up the reserve requirement into the more expensive faster-acting requirement of about ±75 MW, with the remaining ±345 MW coming from slower-responding and less expensive following reserves.

It is often useful to further divide the slower 'following' reserve requirement described above into 'imbalance' and 'variability' components. Imbalance represents the average bias over the time period and is introduced solely due to the inaccuracy of the schedule. This can be an important quantity to ferret out because the imbalance reserve requirement may be reduced through investments in improved wind or load forecasting methods and technology. The imbalance portion of the reserves can be found by substituting 'perfect' schedules—schedules for the operating period that are computed from the average of the observed values—and re-computing the reserve requirements.

Sorting through all the various categories of reserve requirements—for wind versus load, faster or slower response, effect of uncertainty versus variability—can be confusing. Equation (6.1) lays out the foundation for decomposing the various reserve requirement components.[6]

[6] This discussion assumes that all reserve requirement arises from the differences between the expected and actual values of wind and load over the operating period. Other generators may also deviate from the levels demanded of them by operators. These are treated as not significant in this analysis. Large outages that occur with some regularity are dealt with separately through contingency reserves and are not relevant here.

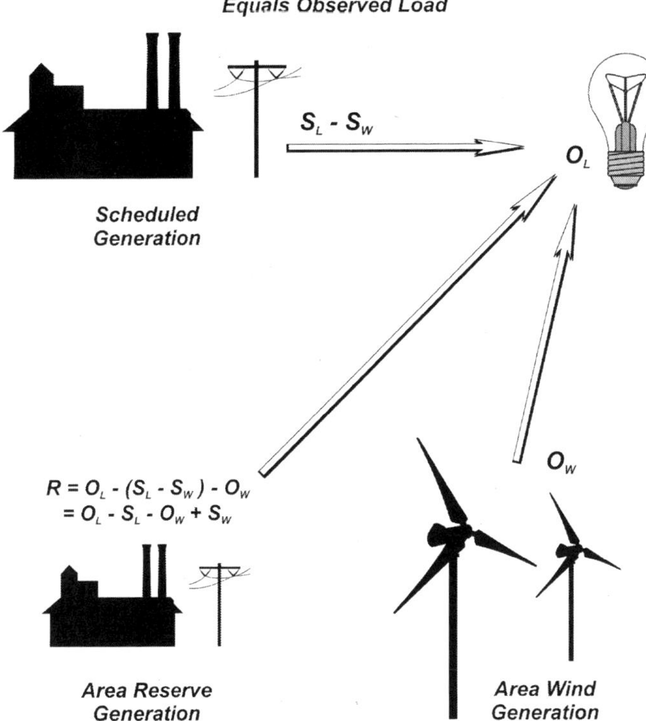

FIGURE 6.9 The reserve dispatch necessary to meet an observed load (O_L) from observed wind generation O_W and scheduled generation for the operating period. Scheduled generation is assumed to be the difference between scheduled load (S_L) and scheduled wind generation (S_W).

Reserves are dispatched to cover the difference between the scheduled wind and load and the observed level of wind and load for any particular time. Figure 6.9 illustrates the basis for the equation. Using the convention that a positive reserve dispatch (R) implies an increase in reserve unit output:

$$R(t) = O(t)_{\text{Load}} - S(t)_{\text{Load}} - O(t)_{\text{Wind}} + S(t)_{\text{Wind}} \tag{6.1}$$

where $R(t)$ is the reserve unit dispatch for time t, $S(t)_{\text{Load}}$ is the scheduled load for time t, $O(t)_{\text{Load}}$ is the observed load at time t, $S(t)_{\text{Wind}}$ is the scheduled wind generation for time t, and $O(t)_{\text{Wind}}$ is the observed wind generation at time t.

The quantities S and O can be subdivided further:

$$S(t) = S(t)_{\text{Perfect}} + S(t)_{\text{Bias}} \tag{6.2}$$

$$O(t) = O(t)_{\text{OPAvg}} + O(t)_{\text{10MinDelta}} + O(t)_{\text{1MinDelta}} \tag{6.3}$$

where $S(t)$ is the load or wind schedule, $S(t)_{\text{Perfect}}$ is a perfect load or wind schedule, $S(t)_{\text{Bias}}$ is the difference between a perfect load or wind schedule and the actual schedule for time t, $O(t)_{\text{OPAvg}}$ is the average of the observed wind or load over the operating period containing time t, $O(t)_{\text{10MinDelta}}$ is the 10-minute average deviation between the observed values and $O(t)_{\text{OPAvg}}$ at time t, and $O(t)_{\text{1MinDelta}}$ is the difference between the 10-minute clock average including time t and the observed wind or load and time t.

Putting all these pieces together, the reserve dispatch at time t becomes:

$$R(t) = \sum_{\text{wind\&load}} [O(t)_{\text{OPAvg}} + O(t)_{\text{10MinDelta}} + O(t)_{\text{1MinDelta}} - S(t)_{\text{Perfect}} - S(t)_{\text{Bias}}] \quad (6.4)$$ [7]

Two of these terms are equivalent—by definition, a perfect schedule reflects the observed values averaged over the operating period,[8] consequently $S(t)_{\text{Perfect}} = O(t)_{\text{OPAvg}}$. Equation (6.4) can therefore be simplified:

$$R(t) = \sum_{\text{wind\&load}} [O(t)_{\text{10MinDelta}} + O(t)_{\text{1MinDelta}} - S(t)_{\text{Bias}}] \quad (6.5)$$

Equation (6.5) contains a total of six constituents, three for wind and three for load. Each term represents the need for a specific reserve dispatch component. The first term in equation (6.5) reflects the variability of the observed values (wind or load) over following reserve timescale (10 minutes in this case), without that needed due to inaccuracy of the schedules. The second term represents the fast changes (minute to minute in this case) in operational need normally covered by regulating reserves. The third term is the reserve dispatch required due to schedule inaccuracy.

Taking a collection of these reserve dispatches for a large number of values of t results in a range of values that constitutes a distribution of values for each of the terms. Each distribution can be characterized by a mean, variance, 95th percentile, etc. If the constituent terms are

---

[7] For simplicity, equation (6.4) is written as a sum over wind and load—we will keep in mind that equation (6.1) is actually the difference between the load and wind terms. This shorthand does not change the important result in equation (6.6), but makes the equations much more compact to write down.

[8] It is possible for these values to be slightly different—schedules usually change linearly (they are 'ramped in') from one operating period to the next instead of changing instantaneously between periods. However, the differences are small and are neglected here for simplicity.

uncorrelated with one another,[9] the variance of the reserve dispatches for many times selected for t can be represented as:

$$\sigma_R^2 = \sum_{\text{wind\&load}} [\sigma_{\text{Schedule Bias}}^2 + \sigma_{\text{10Min}}^2 + \sigma_{\text{Regulation}}^2] \tag{6.6}$$

where σ_R^2 is the variance of the reserve dispatch levels, $\sigma_{\text{Schedule Bias}}^2$ is the variance of the $S(t)_{\text{Bias}}$ values in equation (6.5), σ_{10Min}^2 is the variance of the $O(t)_{\text{10MinDelta}}$ values in equation (6.5), and $\sigma_{\text{Regulation}}^2$ is the variance of the $O(t)_{\text{1MinDelta}}$ values in equation (6.5).

In practice, correlations among the variables are not identically zero, and equation (6.6) is an approximation of the variance of the reserve distribution. Nevertheless, the relatively small correlations make the simple formulation in equation (6.6) both accurate and useful for understanding the relationships among the variables and the reserve requirement.

Reserve requirements are typically set to cover some percentile of the reserve dispatch levels represented in equation (6.5). If those reserve levels follow a normal (Gaussian) distribution, the percentile level is completely determined by the standard deviation of the distribution. For example, 95% of the values in a normal distribution always lie within 1.96σ of the mean of the distribution. While it is not reasonable to expect the distributions represented in equation (6.5) to be normally distributed, the Central Limit Theorem suggests that sums of a large number of random processes will approximate a normal distribution. It is therefore not unreasonable to expect the percentile levels to be proportional to the standard deviation of the distribution, although one cannot in general expect the 95th percentile to lie within 1.96σ of the mean. What this means in practice is that the reserve requirement may be expected to be proportional to the square root of the variance depicted in equation (6.6):

$$RR \propto \sqrt{\sum_{\text{wind\&load}} [\sigma_{\text{Schedule Bias}}^2 + \sigma_{\text{10Min}}^2 + \sigma_{\text{Regulation}}^2]} \tag{6.7}$$

where RR is the total non-contingency operating reserve requirement for both wind and load. Table 6.1 contains load and wind standard deviation data for sample 1-minute wind and load data consistent with the foregoing definitions. The data represent 1 month of data from the Bonneville Power Administration, with system loads averaging about 5900 MW and maximum wind generation of just over 1900 MW.

[9] The formulation in equation (6.6) becomes more complicated if there are correlations among the terms in equation (6.5). The significant correlations (between schedules and observed values) have been removed in defining the variables as they were done here. It is possible, though probably unlikely, for significant correlations to exist between load and wind for wind with a strong diurnal pattern.

TABLE 6.1 Sample Standard Deviations of Distributions Taken from 1 Month of Wind and Load Data

		Schedule Bias	Following	Regulation	Total	Total Estimated by Equation (6.6)
	Load	117.6	60.3	19.7	133.6	133.6
	Wind	140.0	43.1	10.3	145.4	146.8
	Joint wind and load	183.2	72.2	22.1	198.1	198.5
95th	Load	[−268, 206]	[−138, 137]	[−39, 37]	[−246, 294]	
	Wind	[−314, 269]	[−96, 93]	[−21, 21]	[−280, 324]	
	Joint wind and load	[−362, 374]	[−161, 155]	[−43, 42]	[−394, 404]	

'Following' was taken to be the differences between schedules and 10-minute averages of 1-minute data. All values are in megawatts.

The last column of Table 6.1 compares the estimated values derived from equation (6.6) to the individual components, showing the relative accuracy of the approximate formulation and reflecting the low correlations among the distributions examined. Note that, for this data set, the major contributor to the need for holding reserves is associated with schedule bias—a function of forecast (wind or load) accuracy. This is not an unusual result, and is very important in pointing up the value of investments in better schedule accuracy and the cost of holding additional reserves.

One useful purpose of equation (6.6) is in estimating the effects of a reduction in one or more of the constituent components. For example, a 20% reduction in wind scheduling error, dropping the wind bias standard deviation from 140 to 126 MW can be estimated through equation (6.6) to decrease the total joint wind and load standard deviation by about 4.8%. At the 97.5 percentile level of reserves, this corresponds to about a 19 MW reduction in reserves for this data set—potentially worth several million dollars per year; such cost savings likely represent investment worth considering.

Another useful result of equation (6.6) is in attributing the relative contributions of the three contributing components to the need for holding reserves: schedule bias (forecast accuracy), regulating, and following variability. For example, the data in Table 6.1 suggest that schedule bias contributes approximately 85% of the total variance ($183.2^2/198.5^2 = 0.85$). Similarly, the contributions of each of the components can be approximated, potentially for rate-setting purposes (Holtinen et al., 2008).

6.3.2 Conditional reserve requirements

Equation (6.7) is a general relationship expressing reserve requirements for distributions over a range of values for t in equation (6.5). Presumably, the range covers an historical time period of importance—perhaps a year or a month. However, there is nothing in the derivation of these equations to suggest that the time periods cannot be further distinguished. For example, the equations might be applied separately over seasons of the year, and possibly discover that the reserve requirements are significantly different (Holtinen et al., 2008). Another possibility is to prepare separate distributions by hour of the day—potentially leading to time of day-specific reserve requirements. For example, the following reserve requirement tends to be highest during morning and evening load ramp times. Figure 6.10 illustrates that effect over a morning load ramp from about 05:00 to 08:00 when the 10-minute reserve requirement is greatest. These approaches can also be useful for systems where there are strong seasonal or diurnal shapes to the wind resource.

Another possibility is to hold different levels of reserve based on a current or forecast level of wind generation. As a simple example, consider the reserve requirement at times when the wind generation is at

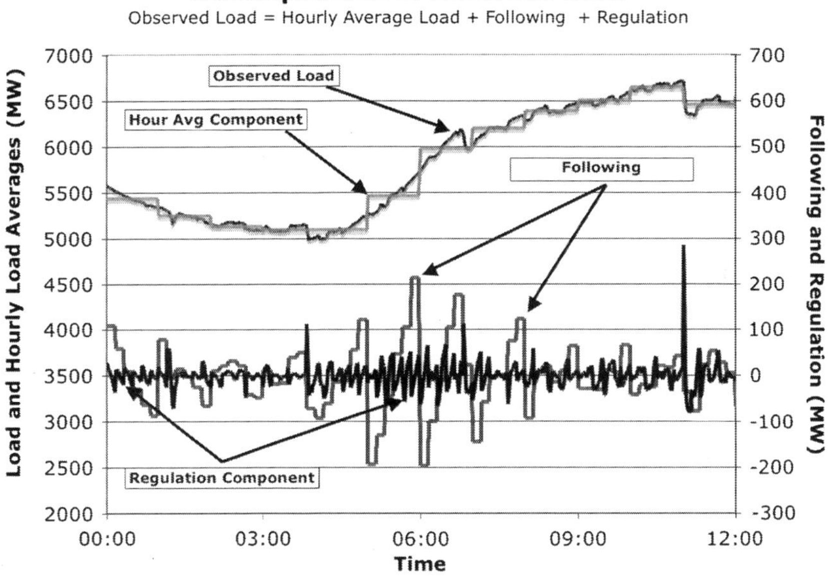

FIGURE 6.10 The constituent components of the observed load formed of the sum of the operating period (1 hour here) average, a following component consisting of 10-minute average deviations from the operating period average, and the regulation component comprehending the remaining minute-to-minute variability. Observed wind generation is similarly decomposed into hour-average, following, and regulation components.

or near zero. When the wind is calm, there is no need to hold significant amounts of incremental generating reserves—i.e. the ability to increase generation should the wind suddenly fall off. If the wind is already at zero, it cannot fall further and there is consequently no concomitant need to hold incremental reserves. Figure 6.11 illustrates the dependence of schedule bias uncertainty as a function of the current production level of wind generation found in one utility wind integration study.

6.4 SUMMARY

One of the key components to determining costs associated with accommodating wind variability and uncertainty is the additional reserve requirements needed to maintain reliability standards. A variety of terms are used to describe both the characteristics of generating equipment providing the reserves, and the base cause for needing to hold reserves. Care must be taken in building the analysis and in communicating with others exactly what is meant without assuming consistency in language.

The purpose of holding and operating reserve generation is to maintain power system reliability standards. Any analysis must be designed in such a way as to determine reserve requirement levels consistent with specific reliability objectives. In general, the analysis will entail developing distributions of levels of reserves needed over historical

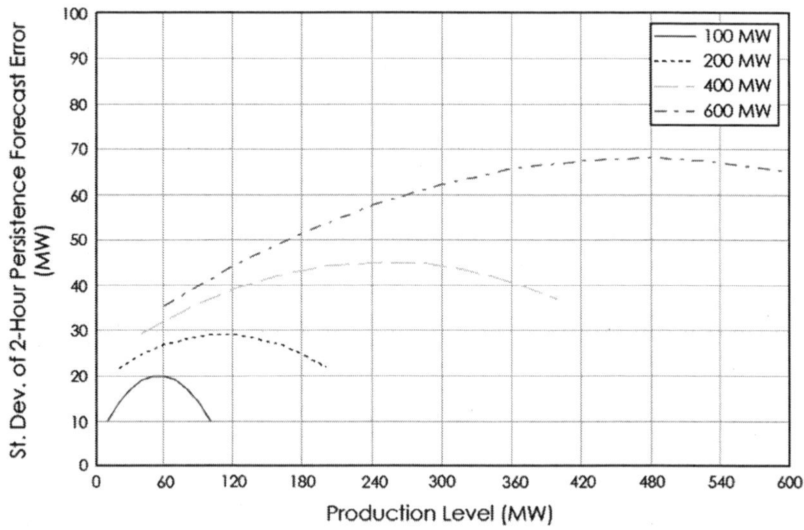

FIGURE 6.11 The dependence of schedule accuracy, here represented as standard deviation of wind schedules based on 2-hour persistence forecasts. The data suggest that the reserve requirement tends to be greatest when wind is currently producing at the midpoint of its generating capability and lowest at both minimum and maximum capability. *Taken from Avista Corporation Wind Integration Study, prepared by EnerNex Corporation.*

or quasi-historical time periods (i.e. estimated behavior of future generation over historical time periods), and picking levels consistent with targeted reliability levels.

Targeted reliability levels are generally proportional to the standard deviation of the reserve requirement distribution. The standard deviation can be broken down into components that separately reflect contributions from load and wind uncertainty and variability. The variability component can be further disaggregated into two categories: the slower-responding following reserves and the faster-responding regulating reserves. The reserve requirement is ideally computed from data in 1-minute increments, but 5- or 10-minute data may be sufficient as the reserve requirements on shorter timescales tend to be modest and can be approximated separately.

Total load and wind reserve requirement is roughly proportional to the square root of the sum of the variances of the individual. Equation (6.7) represents that relationship and can be used to assess the incremental value of improved schedule accuracy or the effects of reduced operating period lengths that would reduce the within-operating period variability. These are key results in assessing the cost of wind generation on power systems.

Finally, the level of reserve requirement need not be considered a static value, but may be adjusted operationally by season of the year, hour of the day, or state of wind generation. Such techniques serve to minimize the cost of holding reserves, releasing reserve units for other purposes when they are not likely to be needed to provide balancing services to wind and load.

REFERENCE

Holtinen, H., Milligan, M., Kirby, B., Acker, T., Neimane, V., & Molinski, T. (2008). Using standard deviation as a measure of increased operation reserve requirement for wind power. *Wind Engineering, 32*(4).

Chapter seven

Wind Power Forecasting

Our best built certainties are but sand-houses and subject to damage from any wind of doubt that blows.

Mark Twain

Chapter 6 laid out the importance of wind forecasting with respect to the need for holding reserves. Wind forecasts play another important role in limiting wind integration costs with respect to unit-commitment decisions that is explored in Chapter 8. It should be clear that the cost of integrating wind is strongly dependent on the accuracy of available wind forecasts. Analysts tasked with determining the value of wind on power systems may not need to become experts in forecasting wind. Nevertheless, costs associated with integrating wind on the power system are at least partly dependent on the accuracy of wind forecasts, and analysts need to understand some of the basics in addition to potentially seeking professional support. Assumptions about forecast accuracy need to be made in the process of evaluating costs associated with accommodating wind on power systems. This chapter lays out some basics about wind power forecasting, the range of accuracy of such forecasts, and prospects for future improvements.

7.1 TYPES AND USES OF WIND FORECASTS[1]

Forecasts of wind generation are produced for a variety of purposes, using a range of techniques. Typically the first forecasts produced for specific wind sites are designed to capture the long-term average behavior for the purpose of determining the economics of the site. On-site measurements may be taken and compared with nearby weather stations with long-term databases that may be statistically correlated with the site. Taking the correlation into account, the long-term average wind speed can be determined, and an estimate of average power generation from the average

[1] See Botterud et al. (2009) for an overview of current practices with respect to wind forecast use in power system and electricity market operations.

Valuing Wind Generation on Integrated Power Systems. DOI: 10.1016/B978-0-8155-2047-4.10007-9

wind speed. Other techniques may also be employed, including the use of physics-based numerical weather prediction (NWP) models, capable of estimating wind behavior at relatively distant points from measuring stations for which long-term records are available. These topics were explored more fully in Chapter 4.

For the purposes of determining costs associated with the dynamic behavior of wind, synthetic operational forecasts are necessary. Long-term average wind generation is an important ingredient in calculating the energy and value of wind from a proposed site, but nearer term, synthetic operational forecasts are vital to determining reserve requirements, an important cost component in accommodating wind generation. There are three basic needs that such forecasts fulfill:

1. Estimating reserve requirements stemming from forecast errors.
2. Short-term (less than 2 hours) marketing or operational decisions.
3. Medium-term (from a few hours to a week or two) unit-commitment decisions.

The effect of forecast errors on reserve requirements was considered in Chapter 6. Costs associated with forecast error through marketing and operational decisions are typically determined using dispatch models. That process is covered in some detail in Chapter 8. Suffice it to say here that power system operators prospectively plan to balance generation and demand over the next hour or two by increasing or decreasing generation or market purchases. They will tend to choose the most economical mix of actions given the information at hand and practical limitations of market liquidity and market transaction time requirements. To the extent wind generation forecasts are in error, their actions will likely be less optimal and may entail greater costs.

Similar plans are made in the medium term for utilities that may have thermal resources taking many hours or even days to fully start up or shut down. Wind forecasts can have an effect on such decisions. For example, if it is known with reasonable certainty that the wind will generate sufficient energy for a period of several days, it may be economical to shut down a power plant for economic or environmental reasons. In North America it is common for natural gas orders to be placed on the day prior to delivery. Another medium-term effect is the extent to which gas contracts may need to be entered into for the next day. These are unit-commitment-related decisions that wind forecasts can affect.

As pointed out in the discussion of producing wind generation in Chapter 4, there is an important distinction to be drawn between wind speed forecasts and wind generation forecasts. Numerical weather prediction models produce wind speed estimates, but those still need to be converted into generation at a specific wind project. Generation levels may be a function of a number of factors that may include wind project power conversion curve, wind-turbine outages, wind direction, humidity, air pressure, wind shear, etc. Even taking into account all these factors, the wind generation across a wind project consisting of many wind turbines will not normally be a single-valued function of wind speed and the other factors. Nevertheless, at least some of these factors should be taken into account in converting a wind speed forecast to wind generation forecasts.

7.2 CLIMATE AND WEATHER

Weather refers to the dynamic state of the earth's atmosphere at any given time. The earth's atmosphere itself is a thin, tenuous layer of gas surrounding the planet, densest near the surface and rapidly declining in pressure and density with elevation. There is no clear demarcation between the earth's atmosphere and interplanetary space. However, 90% of the mass of the atmosphere lies within 16 km, equivalent to only 0.25% of the earth's radius. Weather phenomena familiar to most people are contained within that region, called the troposphere. In proportion, the thickness of the troposphere to the size of the earth is about the same as the thickness of three sheets of paper wrapped around a soccer ball. So thin is the atmosphere that outer space itself is considered to begin just 100 km off the earth's surface, or about the thickness of a coin compared to the soccer ball.

Mount Everest is halfway to the edge of the troposphere at 8.8 km tall, and the air density at that height is half the density at sea level. In contrast to the rapid changes in density, pressure, and temperature that occur in the vertical direction, changes across the relatively expansive surface of the earth are much more gradual. For example, sea level pressure differences of just 10% would be considered very extreme weather events.

The first several hundred meters above the earth's surface are called the atmospheric boundary layer; this represents that part of the atmosphere most affected by interactions with surface geography. Above the atmospheric boundary layer, the atmosphere is relatively unaffected by surface irregularities and is more free flowing, less turbulent. Wind projects exist almost exclusively in the atmospheric boundary layer and, as a result, project output is highly affected by the surrounding terrain.

Sensitive dependence of wind speed, direction, and turbulence on surface features makes forecasts especially difficult. Physics-based weather prediction models must be fine-tuned for each specific wind project location in order to achieve the best results. Moreover, the granularity of such models (typically grid sizes of the order of 2–5 km) is too great to capture the complexity of the terrain in the immediate vicinity of specific wind projects. Various methods are employed to address these challenges, but the basic problem remains.

Although large masses of air can move relatively uniformly at times, differences in wind speeds and direction can be significant over distances as small as meters and centimeters. It is not rare for the wind to be energetic enough to produce appreciable quantities of power in one turbine or set of turbines, while a few hundred meters away there is insufficient wind to energize a modern wind turbine. By the same token, wind also acts on scales as large as hundreds of kilometers in large-scale weather systems such as hurricanes and cyclones.

Climate refers to averages of atmospheric statistics such as temperature, wind speed, and precipitation for a region over decades or centuries. Climate change is rarely, but perhaps increasingly, discussed in the context of wind power. Historically, average wind speeds, precipitation, temperatures, etc. are relatively constant over the centuries. However, it is increasingly clear that the climate is changing. Average temperatures are rising at an alarming rate and effects on weather are noticeable. It may be expected that average wind speeds derived from long-term weather station data may at some point be adjusted to account for expected climate change. At this point this does not seem to be standard practice, either because the changes are expected to be small over the economic life of wind projects, or because the extent and direction of any change is not known with appreciable precision.

7.3 FORECASTING TECHNIQUES

Weather is a result of physical processes that are fairly well understood. A set of mathematical equations governs the meteorological variables: temperature, pressure, moisture, air density, and wind velocity (speed and direction). In theory, these equations can take an initial set of values for each of the meteorological variables at every point in the atmosphere and on the surface of the earth and extrapolate into the future. However, there are not analytic solutions to the equations, meaning that they must be solved numerically, piecemeal in small increments of time and space by large computer models. There are practical limitations that interfere with the accuracy of real-world forecasts due both to the expense and difficulty of undertaking the computational solution to the equations, and to the central complexity of the problem itself.

Chief among the limitations is the completeness and accuracy of the initial conditions. It is not physically (or economically) feasible to measure the state of the meteorological variables at every point in the atmosphere at any one time. Weather station and satellite data are becoming ever more pervasive and have resulted in vast improvements in weather forecast accuracy over time. However, data remain relatively sparse, especially on finer scales in the immediate vicinity of wind-turbine sites.

Compounding the problem of sparseness and accuracy of meteorological data collection was the discovery in 1960 by Edward Lorenz (Gleick, 2008) that the equations governing meteorological conditions were extremely sensitive to the initial conditions—the so-called 'butterfly effect'. Lorenz found that even infinitesimal differences in specifying current conditions can translate to potentially very large changes in the forecast. This sensitive dependence ultimately limits the theoretical accuracy of forecasts based on physical models. Nevertheless, the theoretical limit of physics-based NWP models has not yet been reached and incremental refinements in data completeness and the fineness of detail (geographic and temporal) continue to contribute to forecast accuracy. Due to computational time constraints, NWP models generally divide up the surface of the earth into grid sizes of 2–5 km on a side, and time increments on the order of seconds to minutes. As computing power continues to improve, it is likely that these numbers will continue to fall, with the increased fineness of detail leading to improved forecast accuracy.

Recognition of the sensitive dependence on initial atmospheric conditions led meteorologists to running models over a given forecast period multiple times—observing how the forecast is affected when the assumed initial conditions vary by small amounts. These multiple runs, called 'ensemble forecasts', are used to develop probability distributions for the forecasts. Ensemble forecasts are used in advanced application such as for estimating landfalls for hurricanes to warn populations along the range of uncertainty of the potential danger. Whereas no individual forecast will be completely accurate, the ensemble gives the likely range of possibilities. Such probability distributions may be especially useful to power system operators making decisions hours or days in advance. There are also times where the uncertainties are small and the outcomes of the various members of the ensemble resemble one another. This is equally important information to power system operators.

Physics-based numerical weather prediction models normally take several hours to run and can become stale for the purpose of forecasting the next few minutes or hours. Statistical models provide significant value, particularly on timescales of minutes up to a few hours. Forecast accuracy on these timescales determines reserve requirements and is thus very important to wind valuation analyses. The simplest statistical model is the persistence forecast. In a persistence forecast, the wind generation level[2] is assumed to remain at the last measured value. For example, if wind output at noon is measured to be 100 MW, the forecast for time periods succeeding noon is 100 MW. Variations on the basic persistence model also exist. For example, it may be more accurate to produce a forecast from

[2] Persistence modeling can also be applied to wind speed, but the direct application to wind generation is preferable, as the wind generation over multiple turbines tends to show more persistence than the wind speed at a particular point.

a weighted average of several previously measured values, presumably giving the greatest weight to the most recently measured values, and successively lower weights for older values.

Wind generation levels do, of course, change through time and the further into the future a persistence forecast is used, the greater the error. At some future time, the available measurements are less accurate than simply substituting the long-term average expected generation from a wind project. This suggests another potential refinement to the persistence forecast (Nielsen et al., 1998), which adds a weight for the long-term average that is higher for forecasts into the distant future and lower for forecasts of the immediate future.

Improvements on the basic persistence forecast by weighting with the longer-term expected generation suggest that weighting with the results of an NWP forecast could be better yet. Other methods of combining NWP forecasts with real-time measurements to improve on persistence exist (Ernst, 2005). Statistical models using so-called 'learning algorithms' such as neural networks can provide additional assistance, especially when augmented by meteorological data measurements beyond wind speed (e. g. wind direction, pressure, humidity, time of day, and season) from the local site and surrounding areas.

Another approach that appears promising to improving on the basic persistence forecast is to employ professional meteorologists who interpret the results from NWP ensemble forecasts in view of continuous real-time data available from existing meteorological stations and wind projects. Significant forecast errors are often associated with rapid changes that may occur with the passage of storm fronts. The NWP models may recognize that a front is passing through, but may get the timing off by several hours. Meteorologists may be able to recognize a more or less rapid development of the fronts identified and be able to adjust the forecasts accordingly.

In general, NWP models fitted with modules to translate wind velocity to power generation are more accurate than persistence-based forecasts out past 4–6 hours—the timeframe relevant to unit-commitment and day-ahead marketing decisions. Persistence or persistence–NWP–statistical hybrid models are most accurate for near-term operational decision points. Analysts will have to be prepared to make assumptions about the accuracy of such forecasts in performing wind integration valuation studies.

7.4 FORECAST ERROR MEASURES

Before delving into wind forecast technology, some understanding of the measures of forecast accuracy is necessary. Forecast error calculated for specific time periods is the basic building block for each of the several overall error measures. A forecast made at time t, for some future time increment τ (i.e. forecast for time $t + \tau$) is represented as $F_t(\tau)$. The error in that forecast is simply the difference between the forecast value and the observed levels at time τ, represented as $O_t(\tau)$:

$$E_t(\tau) = O_t(\tau) - F_t(\tau) \tag{7.1}$$

Since forecast error will vary, sometimes being relatively large and at other times relatively small, it makes sense to make some kind of statistical average over time as an overall accuracy measure. Forecast bias is the average error in a forecast over some time period:

$$B_\tau(T) = \frac{1}{N} \sum_{t=t_0}^{t_0+N\delta} E_t(\tau) \tag{7.2}$$

where T is a time period over which the average is taken, starting at time t_0 and extending through time $t_0 + N\delta$, with forecasts updated at each time increment δ. Forecast bias is not normally a very useful measure of forecast accuracy alone because it is not affected by the magnitude of the errors, so long as the sum of the errors is near zero. Most forecast models are careful to monitor and adjust if the bias becomes significantly different than zero over time. Bias is a good diagnostic check to ensure there is no systematic forecasting error in the model.

The two most common accuracy measures are the mean absolute error (MAE) and the root mean square error (RMSE). As the name implies, MAE is a simple average of the absolute values (i.e. treating negative values as positive values) over some time period:

$$\mathrm{MAE}_\tau(T) = \frac{1}{N} \sum_{t=t_0}^{t_0+N\delta} |E_t(\tau)| \tag{7.3}$$

Mean absolute error feels more intuitive to most people, representing the average size of the forecast errors irrespective of whether the forecasts are high or low. However, a potential drawback of MAE is its inability to distinguish between a single large error and multiple smaller ones. For power systems the large errors are of greatest concern and a forecast that reduces the outlier high errors at the expense of some increase in the relatively smaller errors would be deemed a good tradeoff. Therefore, the less intuitive RMSE measure is often preferred:

$$\mathrm{RMSE}_\tau(T) = \sqrt{\frac{1}{N} \sum_{t=t_0}^{t_0+N\delta} E_t^2(\tau)} \tag{7.4}$$

This formulation depending on the sum of the square of the errors gives greater weight to the larger errors than is done in MAE. Squaring the error terms in equation (7.4) ensures that RMSE takes only positive values, as does the MAE. Both RMSE and MAE are affected by any bias in the forecast model. A fourth accuracy measure is the standard deviation of the errors (SDE), which is the familiar statistical concept:

$$\mathrm{SDE}_\tau(T) = \sqrt{\frac{1}{N}\sum_{t=t_0}^{t_0+N\delta}(E_t(\tau) - B_\tau(T))^2} \tag{7.5}$$

This formulation is identical to RMSE in the special case where the bias is zero.[3] The value of SDE is unaffected by forecast bias. Figure 7.1 shows the various accuracy measures as applied to 1 year of persistence forecasts made for they 135 MW Judith Gap wind project in Montana.

The error measures RMSE, MAE, and SDE have 'normalized' versions—NRMSE, NMAE, and NSDE—which are simply the original measures divided by the wind nameplate capacity to form a percentage. The same data represented in Figure 7.1 are repeated in Figure 7.2 in terms of normalized error measures.

Persistence forecasts are often used as bases for comparison of the more complex techniques for short-term (less than 6 hours) forecasts. Any proposed short-term forecast must be more accurate than the simple persistence method. At present, improvements over persistence models are fairly modest up to 6 hours into the future (AESO, 2008). Further improvements are likely with more investment in data collection and analysis.

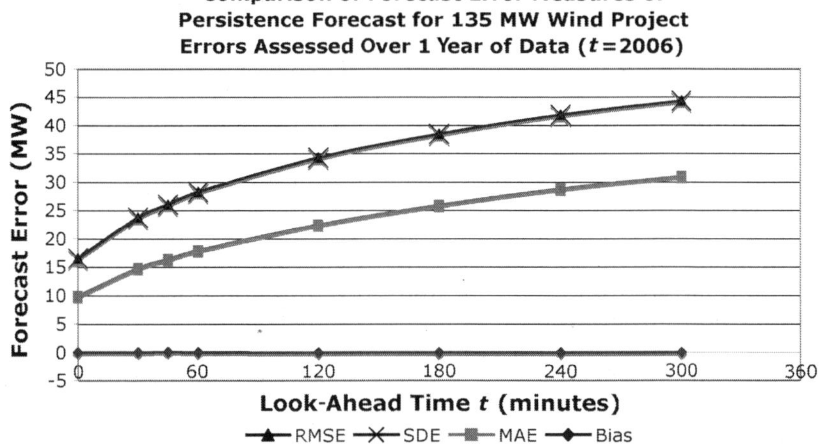

FIGURE 7.1 Forecast accuracy measures applied to one year of simple persistence forecasts of hourly wind generation at the 135 MW Judith Gap wind project in Montana. Note that the SDE and RMSE measures are virtually equivalent because the forecast bias is essentially zero.

---

[3] Forecasts are generally formulated to keep the bias near zero; however, as a practical matter, bias is rarely identically zero.

7.5 FORECAST ACCURACY

Wind power forecasts vary in accuracy depending on a number of factors, including the number of and distribution of the wind turbines being forecast, and the complexity of the terrain in which they are located (Kariniotakis et al., 2004). The Judith Gap wind project data illustrated in Figures 7.1 and 7.2 are exemplary of a relatively large (90-turbine) wind project in complex terrain. Forecasts for collections of wind projects over larger areas will tend to be more accurate than forecasts for smaller projects, or those confined to similar geographic and climactic regions. Figure 7.3 shows the same error measures as in Figure 7.2 over a much larger (≈ 1300 MW) set of wind projects, illustrating the reduction in normalized error as the number of wind turbines becomes large.

The large reduction in normalized error between Figure 7.3 and Figure 7.2 is a result of the relatively weak correlation among forecast errors from distant projects on an hourly timescale. Forecast error correlations normally drop off rapidly over distances of several tens, and certainly hundreds, of kilometers (Fox et al., 2007). In the special case of no correlation among wind project forecast errors, variances of forecast errors would sum to the total variance:

$$\mathrm{SDE}_{\tau}^{2}(T) = \sum_{i=1}^{n} sde_{\tau,i}^{2}(T) \tag{7.6}$$

where $sde_{\tau,i}^{2}(T)$ is the forecast error variance (i.e. square of the error standard deviation) for each individual wind project i, and $\mathrm{SDE}_{\tau}^{2}(T)$ is the variance of the combined forecast error for all the n projects. For a system of uncorrelated wind projects of similar sizes and forecast errors, the

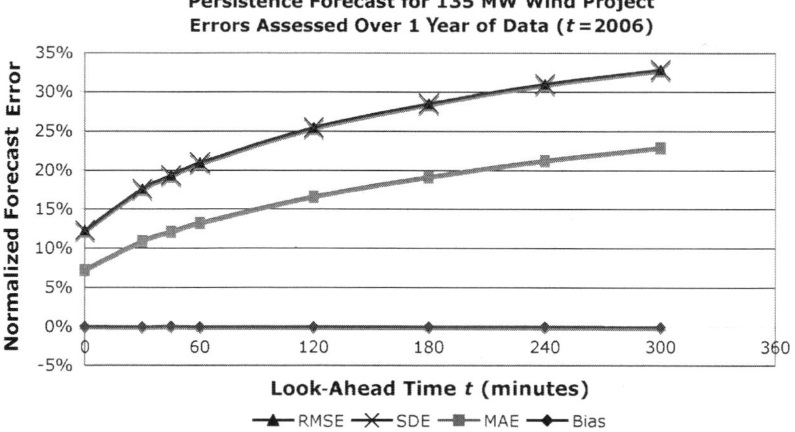

FIGURE 7.2 Normalized forecast errors from the Judith Gap wind project in Montana. Values are identical to those illustrated in Figure 7.1 after dividing by the 135 MW nameplate capacity of the Judith Gap project.

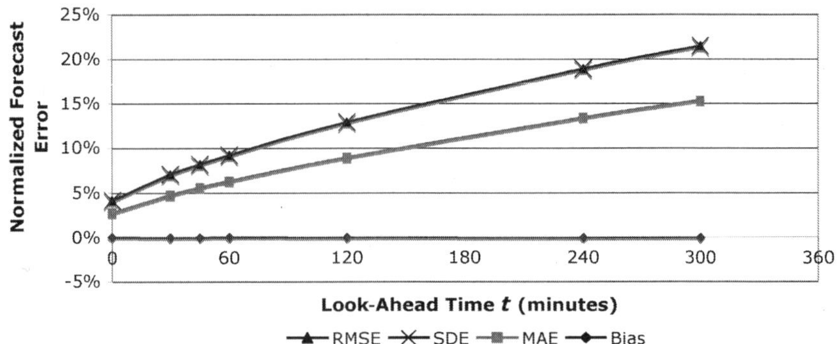

FIGURE 7.3 Normalized persistence forecast errors for a large system of wind turbines in the Pacific Northwest (USA). Note that these values are roughly half of those in Figure 7.2.

normalized $\mathrm{SDE}_\tau(T)$ depends on the number of wind projects in the following way:

$$\mathrm{NSDE}_\tau(T) = nsde_\tau(T)\sqrt{\frac{1}{n}} \tag{7.7}$$

where $nsde_\tau(T)$ is the normalized standard deviation of forecast error for one project and n is the total number of similar projects. This function is illustrated in Figure 7.4, showing the rapid decline in normalized SDE with number of wind projects.

It is important to keep in mind that although the normalized forecast error generally declines with additional wind projects, the absolute error (i.e. in megawatts) actually increases.[4] Even for the special case of uncorrelated forecast errors, SDE and RMSE would be expected to grow with the square root of the number of wind projects.

The special case discussed above represents an idealized circumstance in which forecast errors exhibit zero correlation. Wind projects are often co-located, have different sizes, and their NSDE values vary from project to project. Nevertheless, normalized error will invariably decline with increased numbers of turbine and wind projects except in the extraordinary case where all the forecast errors are highly correlated with one another.

While the behavior of forecast errors shown in Figures 7.2 and 7.3 represents typical results for simple persistence forecasts of terrestrial wind

[4] Forecast error in absolute terms would only decline with additional wind projects if the additional forecast errors were for some reason negatively correlated with the aggregate of the other forecast errors. While forecast errors might be positively correlated for physically nearby wind projects, negatively correlated errors are very unlikely.

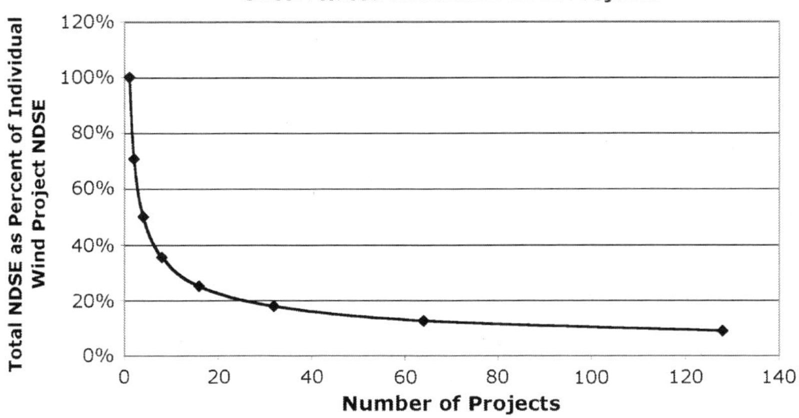

FIGURE 7.4 The rapid decline in normalized standard error in forecasts for the special case of uncorrelated forecast errors. Note that although, as illustrated, the normalized standard deviation declines, the standard deviation in absolute terms grows with the square root of the number of projects. In addition, this curve is an effective lower bound on actual relationships where forecast errors exhibit some degree of correlation.

turbine projects, they are not representative of NWP models or hybrid persistence models. Forecast accuracy improvements made by combining NWP and persistence, or adding in other statistical models, tends to flatten out the curves in Figures 7.2 and 7.3, increasing their accuracy at the longer time horizons (Kariniotakis et al., 2004). Forecast horizons in excess of about 4 hours almost always benefit from inclusion of forecast information from NWP models. Numerical weather model accuracy becomes most relevant in the 4-hour to several-day timeframe. Past about 7 days, the climatological record becomes more accurate than weather model forecasts.

Just as forecast errors from various wind projects reduce the overall forecast error, forecast errors from competing forecasts will also tend to cancel one another (Lange et al., 2007). Figure 7.5 illustrates the improvement seen by combining three separate day-ahead NWP model forecasts made over a 7-month period for wind projects in Germany. NWP results from different model configurations were fed into a single artificial neural network (ANN) to produce power forecasts for the wind turbines. Combining the three forecasts as a simple average reduced the overall forecast error (NMRSE) by more than 20%.

7.6 DEVELOPING SYNTHETIC FORECASTS

Accurate valuations of wind, with the attendant cost of integrating wind into the power system, depends on developing reasonably credible synthetic wind forecasts to complement the wind data described in

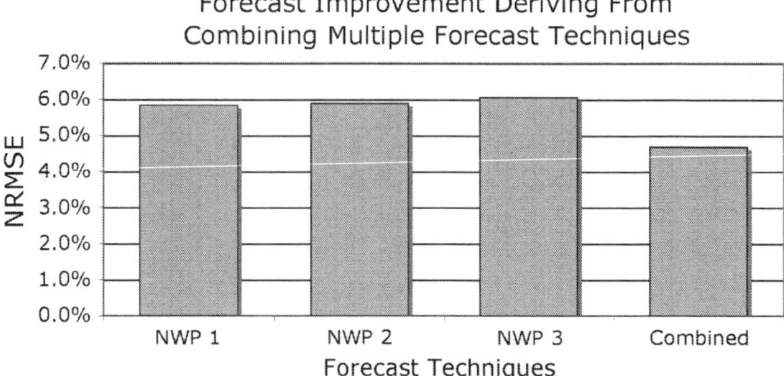

FIGURE 7.5 Forecast improvements realized from combining forecast techniques of multiple numerical weather prediction models. Forecast errors tend to cancel one another, resulting in an overall improvement in forecast accuracy by combining model forecasts. *Adapted from Boyle (2007).*

Chapter 4. For short-term forecasts on which reserve levels are set, a persistence-type forecast is easily produced from the wind power data developed. Longer-term forecasting is somewhat more complex. The day-ahead forecast error for an aggregate system of wind turbines and projects depends on (Lange et al., 2007):

- Individual wind farm characteristics and complexity of local terrain.
- Number of wind turbines and projects.
- Accuracy of the weather prediction model used.
- Accuracy and completeness of model input data.
- Season of the year.

Whereas a reasonable short-term (e.g. hour-ahead) forecast can be easily produced using persistence methods on a synthetic wind data set, this cannot be as directly accomplished for day-ahead forecasts. One method is to take the wind power generation for the forecast day adjusted by a random component to form the forecast. The appropriate size and distribution of the random components is not clear. However, local experience or data such as that shown in Figure 7.5 can give an order of magnitude understanding of the likely forecast error. A further refinement would be to use estimations from numerical weather models based on historical data inputs. This can be used to gauge the range of uncertainty and distribution of longer-term forecast errors.

7.7 SUMMARY

Assessing the cost of integrating wind is importantly connected to assumptions about the accuracy with which the wind is forecast. Short-term forecasts are used to set incremental reserve requirements, and longer-term (typically day-ahead) forecasts affect costs incurred in balancing the system on a planning basis.

Short-term forecasts can be as simple as a persistence forecast in which a current level of wind project output is assumed to remain unchanged through time. More complex statistical-type forecasting techniques exist and may be combined with physics-based numerical weather prediction (NWP) models to improve their accuracy. NWP models are in turn used more commonly for longer-term (e.g. day-ahead) forecasts.

Analysts may produce synthetic short-term forecasts to be used in their analyses using persistence techniques applied to the base wind data developed as discussed in Chapter 4. Simulating day-ahead forecasts is best accomplished using numerical weather models using historical data, although it may be sufficient in some cases to simply add in an error term to the day-ahead base wind data. Such error terms may have to come from an assumed distribution, or may be informed by ranges of outcomes in NWP ensembles.

REFERENCES

Alberta Electric System Operator (AESO), Alberta Energy Research Institute, and Alberta Department of Energy (2008). *Wind Power Forecasting Pilot Project*. Available at: <http://web.ta-alberta.ca/downloads/Work_Group_Paper_Final_(3).pdf>

Botterud, A., Wang, J., Monteiro, C., & Miranda, V. (2009). *Wind power forecasting and electricity market operations*. 32nd IAEE international conference.

Boyle, G. (2007). *Renewable electricity and the grid*. Earthscan.

Ernst, B. (2005). Wind power forecast for German and Danish networks. In T. Ackermann (Ed.), *Wind power in power systems* (pp. 365–381). Wiley.

Fox, B., Flynn, D., Bryans, L., et al. (2007). Wind power forecasting. *In Wind power integration: Connection and system operational impacts*. Institution of Engineering and Technology (IET). (pp. 209–237).

Gleick, J. (2008). The butterfly effect. In *Chaos* (pp. 11–31). Penguin.

Kariniotakis, G., Marti, I., Casa, D., et al. (2004). What performance can be expected by short-term wind power prediction models depending on site characteristics? *Proceedings of the European wind energy conference EWEC*, 22–25 November 2004.

Lange, B., Rohrig, K., Schlogl, F., et al. (2007). Wind power forecasting. In G. Boyle (Ed.), *Renewable electricity and the grid: The challenge of variability* (pp. 96–120). Earthscan.

Nielsen, T. S., Joensen, A., Madsen, H., et al. (1998). A new reference for predicting wind power. *Wind Energy, 1*, 29–34.

Chapter eight

Wind Energy Valuation Studies

The pessimist complains about the wind; the optimist expects it to change; the realist adjusts the sails.

William Arthur Ward, American author, 1921–1994

Economic analyses are performed to examine the cost-effectiveness of wind resources—usually in conjunction with resource planning efforts engaged to determine the best combination of resource additions to meet customers' needs. This chapter reviews the basic changes that analysts may make to commonly available dispatch models necessary to capturing the relatively unique operating characteristics of wind generation.

There are two general approaches to wind-specific economic analyses. In one approach, wind is treated like any other resource, but with characteristics that need to be recognized and accounted for in the analysis (e.g. increased balancing reserve requirements). Studies are designed to treat wind like other resources, while capturing all of the special operating characteristics that are expected to contribute significantly to costs. As discussed in more detail in Chapter 5, appropriately representing wind operating characteristics in existing power system models can be challenging. The second approach is to capture some or all of the special characteristics as a 'wind integration cost' that is calculated separately, and added to the valuation study separately. Chapter 9 is devoted to determining wind-specific costs that may also be used by balancing areas to impose wind integration tariffs.

The value of energy generated by wind power facilities could in principle be determined by multiplying an expected pattern of wind generation through time by a forecast level of market prices for power. This approach would be consistent with the view of an independent wind generator bidding into a liquid wholesale marketplace. In adopting such a view, the resulting revenues would need to recognize certain ancillary service costs that may be imposed by the balancing area authority. Unfortunately, wind-specific ancillary service charges are in a nascent stage

Valuing Wind Generation on Integrated Power Systems. DOI: 10.1016/B978-0-8155-2047-4.10007-9

of development and, where they do exist, may not reflect the actual costs imposed by wind generation very accurately.[1] In addition, where wind is a new or rapidly growing source of energy, it can directly affect the cost of providing energy and ancillary services.

From a power system perspective, a more in-depth analysis is called for that includes the effects on ancillary services. Similarly, extensive wind generation can affect wholesale energy market prices as well. For these reasons, where significant amounts of wind generation are contemplated, it is necessary to derive the value of wind generation from an integrated power system point of view, and conduct more detailed analysis of the effect of wind generation on power systems as a whole—the topic of this chapter.

The more complex analyses are usually performed using chronological economic dispatch models (CEDMs). CEDMs are used to evaluate the costs and benefits of a wide variety of prospective changes affecting power systems, including acquisition of one or more competing generating resources, changes in market conditions, changing environmental requirements, and acquisition or outages of large power system components such as transmission lines or substations. CEDMs emulate much, but not all, of the complexity of modern power system operations and analysts often rely on them to assess the economics of a major policy or physical change to the power system. Most complex wind valuation studies are done using CEDMs due to the complexity of the interaction between the time-varying nature of wind generation, demand for power, and the more flexible generators on the system. These models evaluate power system operations through time, finding the least costly generating resources to meet demand over each time increment examined. They depend on inputs such as the operating costs and characteristics (e.g. maximum and minimum generating limits, maximum output change per minute, minimum down times, startup times, minimum run time, etc.). The primary result reported by CEDM runs is the total cost incurred in meeting power demand over the study horizon.

In performing valuation studies with CEDMs it will be important to understand the limitations of the available model and ensure that the representation of the power system response is realistic. As pointed out in Chapter 5, models may not be able to capture certain nuances in the planning (unit-commitment) process. Care must also be taken in representing the reserve requirements described in Chapter 6. Costs not typically captured by CEDMs are identified here, but explored more fully in

[1] Lack of accuracy stems from both the difficulty of determining the costs of ancillary services for wind, but also the complexity of rate structures that would recover costs in proportion to such difficult-to-measure metrics as variability and schedule accuracy. Tariffs are typically charged on the basis of simpler metrics such as installed capacity or energy generation.

Chapter 9. This chapter will point out the basic steps in preparing inputs to the models and interpreting model output.

8.1 SYSTEM RESPONSES TO WIND GENERATION

The purpose of running wind valuation studies is to capture economic consequences stemming from the power system's response to wind generation that would not otherwise be easy to determine. Analysts need to be clear at the outset what these effects are expected to be, and whether the available model can capture the relevant effects. Among the important economic effects of wind in power systems, CEDMs are used to determine the following:

1. Reduced overall costs to the power system from the presence of zero- or low-variable-cost wind generation.
2. Opportunity costs from holding higher levels of generating reserves to meet the variability of wind generation.
3. Costs associated with wind generation uncertainty, including increased costs from wholesale market purchases, and efficiency losses from higher needs to change output at flexible generating units.
4. Effects on wholesale electric and fossil fuel market risk.

For the purposes of this chapter, economic analyses will be viewed from an overall system cost perspective. However, there are localized effects of wind generation that may be of interest or concern to individual parties. For example, additional wind generation is likely to result in less energy produced by relatively high marginal cost resources on the power system. From the perspective of an individual fossil-fuel generator displaced frequently by wind energy, the economic impact might be high. Such an effect may conversely be deemed a benefit from net purchasers of energy, who find their cost of meeting power demand reduced. It is important to keep in mind the perspective adopted by any particular study. Again, for the present purposes, the economic analyses considered here are those at the system-wide level.

The presence of wind energy tends to reduce operations of other generating units on the power system, reduce purchases from and increase sales into wholesale power markets. CEDMs, with wind appropriately represented, should capture these effects. Wind is typically represented as hourly patterns of generation whose presence allows the model to meet demand using less energy from other sources. Correctly representing wind patterns (see Chapter 5), variable fuel and operating costs of competing resources, and wholesale markets is essential.

8.2 STUDY DESIGN

Wind valuation studies can be organized in pairs—a status quo model run over some study horizon and a second study over the same time period, with the power system changes (e.g. resource additions) that are to be examined. Output of the models is usually summarized as a present value of the operational costs (and potentially some capital costs) associated with the model runs. Chief among the costs and revenues that may change are those associated with sales and purchases in the wholesale power markets, and fuel costs for those resources with substantial variable fuel costs. Adding wind resources will tend to increase sales, decrease purchases, while reducing fuel consumption and emissions from fossil-fueled resources.

The reduced costs and increased revenues realized in the model runs are offset by other costs associated with building and operating wind generators, as well as with accommodating the variability and uncertainty of their output on the balance of the power system. Not all of the costs are computed directly in the models themselves. For example, the cost of building and operating a wind project is treated either as model input, or may simply be accounted for outside the model altogether, taking account of other savings determined in the model runs. In addition, few models are designed to analyze system behavior in time increments smaller than 1 hour. Costs associated with operating reserve units whose operation is specifically tied to within-hour operating needs (regulating reserves) are not typically accounted for within the models themselves. However, most models allow setting aside specified reserve generation for the purpose of maintaining power system reliability. To the extent that the requirements may increase (e.g. due to additional wind on the system), the models will capture the opportunity cost associated with holding that generation aside.

The basic steps in designing a wind valuation study include:

1. Developing wind generation data as a suitable model input.
2. Developing external costs (capital and operation and maintenance (O&M) costs, or contract purchase costs).
3. Determining intra-hour reserve requirements for model input.
4. Running a model with wind and increased reserve requirements. This may require a two-step process:
   a. A run to determine unit commitment based on wind and load forecasts.
   b. A second run with wind generation and load data where unit commitment decisions are fixed from the previous run.
5. A model run or runs with competing resource portfolios.
6. Comparison of the present value costs in the competing model runs.
7. Study validation.

These steps may be repeated with differing assumptions for important parameters such as fuel prices, capital costs, etc., to examine how the relative costs among the competing portfolios may change. This is commonly done for risk analysis purposes to be discussed in fuller detail below.

Dispatch models vary in their ability to handle reserve generation set-asides. At a minimum, a given model will likely allow for reserve generation to be held as a fraction of the load for a given hour. Higher levels of sophistication allow for a combination of percentages and megawatts of generating capacity that can be specified separately by month, year, or hour of the day. Different types of reserves may also be an allowable specification—whether the reserves can be held only on generating units already operating, or a combination of operating and quick-start generating units. Few commercial models will allow a dynamic specification of the reserve requirement by load and wind state—for example, higher levels of reserves over hours with greater wind/load variability, and lower amounts at more stable times. In any case, the opportunity costs associated with holding wind-related reserves will be accounted for if they are represented in the models. Essentially, the model will observe the reserve requirement for each operating period and set aside generation on the (generally) highest variable-cost resources available.

More complex studies can be designed to look at alternative levels of wind generation or combinations of resource additions planned to meet load growth. Multiple studies can be compared using measures such as the net present value of system costs. Each set of resource additions over the study horizon may be referred to as a 'resource portfolio'. Resource portfolio costs can be compared across different future scenarios. Future scenarios usually include different expectations with respect to growth in power demand, and wholesale electric and fossil-fuel market prices. This profusion of portfolios and scenarios can become confusing, but the basic premise is to compare costs of alternative resource choices over a study horizon.

8.3 MODEL MODIFICATIONS FOR WIND

Commonly available dispatch models will need modifications in order to capture costs associated with wind variability and forecast uncertainty. It should be noted that these characteristics are similar to system demand and some other generating resource types (e.g. run-of-river hydro). However, the rapid development of wind generation and relatively low level of familiarity among analysts has resulted in much attention being given to understanding the costs associated with wind variability and uncertainty.

Capturing costs associated with variability generally means increasing the level of reserve generation set aside to balance the system and maintain reliability. Costs associated with uncertainty are captured by supplying the model with forecasts of wind generation used to commit resources or transact in available wholesale markets to reduce the balancing reserve requirements. Often, models will need to be run two (or more) times, fixing unit-commitment and market transactions to the forecast wind, and then again with the wind generation data. Forecast inaccuracy results in greater reliance on following and regulating reserves, and can be an important aspect in the valuation.

8.3.1 Modeling variability

Modeling the effects of intra-hour wind variability is primarily accomplished by causing the model to maintain appropriate levels of reserve-generating capability. The methods used for determining the appropriate reserve levels are described more fully in Chapter 6. Models generally allow for nominating an amount of spinning and non-spinning reserve—usually expressed as a fraction of total generation for a given operating period, but sometimes as specific megawatt amounts as well. Reserves are held for multiple reasons, and the necessary amounts for the different purposes may need to be summed for input to the model. For example, contingency reserves to cover sudden power system component outages will be in addition to the regulation and following reserves covered in Chapter 6. Care must be taken in summing types of reserves to ensure that there is no double count. For example, the regulating reserve requirement stems from the combined effects of wind and load; however, as discussed in Chapter 6 it is not correct to calculate regulating reserve requirements for wind and load in isolation and add the two requirements together to find the total.

Most CEDMs simulate operations in time increments no shorter than 1 hour. Sub-hourly operations of generating units (for regulation, following, and contingencies) are not explicitly modeled. Supplying the models with reserve requirements ensures that the specified levels of generation are not dispatched for other purposes, and are available for the assumed within-hour balancing needs.

For hourly models, it is important to recognize that the inter-hour reserve requirements should *not* be supplied to the model as a reserve requirement on the same basis as regulation, following, and contingency reserves. Because hourly models will attempt to operate available resources to meet imbalances arising from schedule errors, holding those reserves out of the dispatch may cause the model to incorrectly flag reliability events that would not arise in actual operations. The reason for this is that the model algorithm will withhold those reserves and not recognize their availability, and use them as necessary. At least one wind integration analysis was performed that suggested significant increases in failures to meet load paradoxically while simultaneously holding hundreds of megawatts of reserves for the purpose of avoiding just such shortfalls.

It may be possible to change model algorithms to accommodate a new category of 'last resort' reserves to represent inter-hour reserve requirements. Again, this is not different than logic that should be available to accurately represent similar issues with load, but is not commonly available in CEDMs today. In a similar vein, the model should not be allowed to commit all available resources to load or market transactions because they may be needed for balancing purposes should the actual load and wind be significantly different than the expected levels.

8.3.2 Modeling forecast uncertainty

Wind forecast uncertainty enters into the modeling on two different timescales. Both timescales are related to minimizing the amount of relatively high-cost balancing resources required to accommodate wind—those resources withheld as just previously described to accommodate variability.

The first important timescale is of the order of many hours to several days—the time it takes to start up or shut down operations at large (mostly fossil-fueled) steam generating units. This same timescale is relevant to the time it takes to engage in day-ahead (or longer) market transactions. Real-world decisions to start up or shut down power plants are emulated in CEDMs in what is called "unit-commitment logic".

The second timescale is determined by the time it takes to engage in market transactions and generator schedule changes that can also be used for balancing purposes. This timescale is characterized by 30–120 minutes prior to the operating hour.

As a consequence of the natural timescales—day(s) ahead and hour(s) ahead—wind and load forecasts available at those times are important to the economically efficient integration of wind on power systems. Although the effect of demand forecasts is equally important to those of wind, CEDMs have not typically been designed to capture the value of the accuracy of load forecasts. While operators have long experience with the accuracy of load forecasts, the accuracy of wind forecasts is less well known and currently evolving. Indeed, analysts may often assume the value of wind forecasts without having explicitly evaluated (or even understood) how the value of wind energy is explicitly dependent on forecast accuracy.

To be clear, the more accurate the wind forecast, the smaller the level of inter-hour reserves that must be held. At current levels of forecast accuracy (hour-ahead forecasts based on persistence), the majority of incremental reserve requirements for wind are inter-hour reserves due to forecast error. The assumed level of forecast accuracy becomes one of the most important factors in the wind valuation study. Care must be taken to represent forecast accuracy appropriately.

At present, CEDMs are not typically designed to allow users to specify demand and wind forecasts, along with patterns of simulated demand and wind generation. Capturing the interaction may require analysts to feed the model forecasts of wind and load, freeze the resulting unit-commitment decisions based on the forecasts, and then rerun the models with simulated wind and load and releasing reserves held for forecast error for potential dispatch during the modeled operating periods. This process should be duplicated in the with-wind and without-wind model runs.

Studies employing CEDMs over 20-year study horizons, potentially running multiple times for scenarios or stochastic variables, can be relatively time intensive—emulating unit-commitment decisions based on forecasts can become a daunting increase in work for the analyst. One answer to this problem is to run special studies to estimate this effect and merely include the resulting 'integration cost' as an input to the model. Such special studies are the subject of the next chapter. Given the growing importance of wind on modern power systems, CEDM vendors may be encouraged to incorporate options in their products to automate the process of basing unit-commitment decisions on forecasts.

8.4 EXAMPLE STUDY RESULTS

An example of how the steps fit together may be helpful. The figures in Table 8.1 represent the outcome of CEDM studies over a 20-year study horizon for each of three competing resource additions—simple portfolios consisting of alternative resource additions. The first portfolio represents the base system relying on wholesale market purchases to meet 100 MW of additional load. Portfolio 2 is the base portfolio plus 300 MW of wind generation. Finally, the third portfolio constitutes the base system plus a 100-MW combined cycle combustion turbine (CCCT).

The rows of Table 8.1 represent steps in the analysis. The capital cost row is an input to the CEDM, showing the present value (PV) of the capital additions needed for both the wind and CCCT portfolios. Since the base system capital costs are identical for the three portfolios, only the incremental capital costs are relevant.

The table includes operating cost results from the interim unit-commitment study run referred to as step 4(a) in Section 8.2. In this example, it is necessary to run an interim CEDM based on forecasts of load and wind to allow the model to determine which generating units are started up and available at the various times through the study horizon. The 'simulated load' and 'wind runs' shown in the table are the resulting operating costs when the models are rerun with unit-commitment decisions fixed to the results in the previous step, but with simulated load and wind levels. Note that the operating costs are slightly higher in these runs, reflecting the cost associated with load and wind uncertainty.

Overall, the total operating plus capital costs are lower in this example for the wind addition than for the CCCT or market purchase portfolios. The 'net value versus base system' row relates the relative value of the wind and CCCT portfolios in terms of dollars per megawatt hour as another, possibly more familiar, measure of relative value.

TABLE 8.1 An Example Analysis of Three Competing Resource Portfolios*

	Base Power System Costs with 100 MW Market Purchases	Power System Costs with 300 MW Wind Addition	Power System Costs with 100 MW CCCT Addition
Capital cost additions adjusted for tax credits (millions real 2009 $)		442	116
Operating costs unit-commitment run (PV millions 2009 $)	10,512	9810	10,249
Operating costs simulated load and wind data run (PV millions 2009 $)	10,643	9981	10,381
Operating plus capital costs (PV millions 2009 $)	10,643	10,423	10,497
Net value versus base system (levelized $/MWh)	—	12.55	8.38

**A base system without modification purchasing as needed from wholesale markets, the base system plus 300 MW of wind, and the base system plus 100 MW of combined cycle combustion turbine (CCCT) generation. The last row shows the net benefit of the alternatives over the base power system.*

In practice, studies may be more complex than this simple example, consisting of more portfolios, more scenarios about market prices, potential carbon dioxide costs or penalties, and capital cost assumptions. Further, load growth is likely to come on more gradually through time, while generating unit additions may come on at various times. Nevertheless, the simplicity of this example is designed to illustrate the basic steps and how they relate to one another.

8.5 PORTFOLIO RISK AND WIND GENERATION

One of the potentially important value considerations is how highly capital-intensive resources such as wind generation may reduce utility company exposure to wholesale electricity and fossil-fuel price volatility. Studies of costs under various future scenarios result in multiple costs associated with a given resource portfolio. Such studies form the basis of revenue cost and risk tradeoff analysis. For tradeoff purposes, an expected or base level of costs is treated as one objective that is plotted against a risk metric. There is no universally agreed upon risk metric, but a number of measures may be considered and one or more chosen as the basis for the cost–risk tradeoff.

Often, the risk measure will reflect the average of the highest cost scenario outcomes—the so-called expected tail risk. For example, if 100 scenarios are examined for each portfolio, the 10 worst outcomes (highest costs) of those 100 might be averaged as the relevant risk metric. This allows analysts to examine the tradeoff between costs and risks. Irrespective of the multitude of potential risk metrics, this procedure does not objectively result in a least-cost, least-risk portfolio. However, weaker portfolios may be weeded out as being both higher cost and higher risk than other competing portfolios. The remaining 'robust' portfolios exist on an 'efficient frontier' of cost and risk.

Figure 8.1 illustrates a plot of 12 resource portfolios, with expected costs on the horizontal axis and a risk metric on the vertical axis. All but three of the portfolios are weeded out on the basis of being both higher risk and higher cost than at least one of the three efficient frontier portfolios. Portfolios P1, P2, and P3 are the efficient frontiers in this example. As noted above, there is no objective means of choosing a best candidate from among the efficient frontier portfolios. Portfolio P1 is the lowest-cost portfolio, but both portfolios P2 and P3 are lower risk. Decision-makers are still left with the question of whether the reduced risks represented in P2 and P3 are worth the associated higher expected costs. It should be noted that choosing a different risk metric could result in a different efficient frontier.

Historically, fossil-fuel and wholesale electric market prices have exhibited some of the highest volatilities. As a result, the highest risk

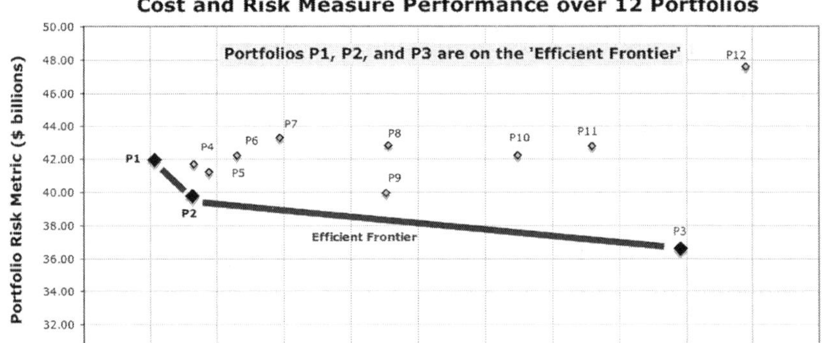

FIGURE 8.1 The cost–risk tradeoff among 12 competing resource portfolios. Efficient frontier portfolios are those for which there does not exist a competing portfolio with both lower costs and lower risk. For example, P9 is not on the efficient frontier because P2 is both lower cost and lower risk. Conversely, P3 is on the efficient frontier because although it has among the highest expected costs, no other portfolio has lower overall risk.

scenarios are generally those associated with high fuel and market prices. Consequently, resource portfolios with lower market purchases and fewer fossil-fuel-dependent resources will have superior risk properties. Renewable resources such as wind, solar, and geothermal fit this profile. Power systems that are heavily dependent on fossil-fueled resources will likely find renewable resources reducing their overall risk.

8.6 COSTS AND VALUE NOT CAPTURED BY CEDMs

Some costs not captured by CEDMs may need to be added into the valuation study separately. The two most likely candidates are the increased operating and maintenance costs on generators that have to operate more frequently and ramp more rapidly, and the reduced overall efficiency of generating units due to operating outside their peak efficiency points.

Wind variability and uncertainty increases not only the amount of generation withheld for balancing (the value of which is generally captured by CEDMs), but also the frequency and rapidity with which generation levels on those units change. Most operators recognize that more extreme dynamic movement of generation levels increases the wear and tear on machinery, and consequently increasing overall operating and maintenance (O&M) costs. Quantifying any such impacts is generally a challenging problem. Often, analysts query operations personnel for estimates of the effects of such increased operations. Responses vary widely and may not be particularly reliable. A more objective approach would quantify dynamic operations (e.g. number of unit startups per month, average rate of change of output while operating, etc.) and relate frequency of repair to such quantities. This is often outside the scope and budget of wind studies, and the increased O&M costs are either very roughly estimated or omitted altogether as not being significant.

Power system operators will operate generators at their peak efficiency curves as much as possible. Reserve units will operate at various output levels around the highest efficiency point. Increasing the dynamic range of output levels for these units to accommodate wind variability means lower overall efficiency for these units. A fuller discussion of this effect is given in the next chapter.

Some studies may accord value to generators based on their contribution to meeting peak demand—sometimes referred to as a capacity credit or capacity payment. Chapter 10 reviews methods used to determine the capacity contribution of wind projects. Where studies capture a capacity credit or value to other generation, it is reasonable to accord such a value to wind generation. In some analyses, minimum reliability criteria must be met without a specific value attached to each individual generator contribution to meeting the criteria. A general guide is to treat wind generation as similarly as possible to other generating technologies.

8.7 STUDY VALIDATION

For large power systems, the amount and complexity of the data that feeds CEDMs can become overwhelming, notwithstanding additional complexity contributed by wind generation. It is imperative to institute quality control procedures on model inputs, changes to algorithms, and model results. Limits on time and staffing resources can make instituting the necessary validation procedures especially challenging. Nevertheless, it is important to develop some procedures to limit errors to the extent practical.

8.7.1 Input validation

Best practices in model input validation entail documenting each change in data, citing sources for the data entered, reasons for the change, date of change, and identity of both the analyst requesting the revision and the person entering the change into the database. Ideally, each data item would be independently reviewed and verified by someone different from the person entering the data. It should go without saying that the units of the data should be explicitly stated in the data entry interface and retained in the database.

Development of automated error checking for inconsistent data or statistically outlying data should also be developed to flag potentially erroneous data. For example, thermal unit heat rates are normally within a range of values and error-checking logic should be instituted that flags any heat rate data entered that is outside that range.

It is all too easy to enter data in the wrong units, or to misplace a decimal point. Alarmingly few data input interfaces provide assistance in detecting such errors. Similarly unusual are serious procedures to document and double-check data inputs to the models. Given the literally tens million of numbers that typically go into these studies, it should be obvious that serious systems for input analysis, including automated error checking, are vital to executing credible CEDM studies.

8.7.2 Algorithm validation procedures

Tight controls on any major model algorithm changes, or data input changes that exercise algorithms not normally engaged, need to be carefully validated. Validation should include careful annotation of programming code, dates and reasons for the changes, documentation of the programmer or analyst making the changes, review by another programmer or analyst, and careful beta testing of the new procedures. Program logic is often used for purposes slightly different than those intended, making full documentation of how the new routines work and what they do critical to the user. This is another area where practice often falls short of the ideal not only for lack of resources, but also because algorithms are often considered proprietary and full disclosure by commercial vendors is sometimes seen as revealing competitive advantages to competitors who may then take advantage of them. This seems particularly true of unit-commitment logic. Analysts may have to spend considerable time running tests to reverse-engineer how the unit-commitment logic behaves, as vendors are often reluctant to make that information available.

8.7.3 Validating results

Ensuring the validity of data inputs and algorithms is crucially important to building confidence in the accuracy of the results of complex studies. Just as important, however, is examining the results of studies for basic reasonableness. Some commercially available models incorporate the output of the models in database management systems similar to, or even part of, the input database system. Building a library of prior results can also be of use in automating validation of output results. Routines can be built to examine the distributions of various results (e.g. capacity factors of existing generating units, loss of load events, and overall revenue requirements, etc.) that can be used to flag unusual results for further investigation. Unusual events should have an understandable explanation (e.g. due to a significant change in market assumptions, load growth, resource additions, etc.), or may signal a modeling error that may arise due to erroneous data, or failures in model logic, or just as likely an unusual combination of both effects that causes an unrealistic simulation of power system operations.

An important validation tool is analyst estimation of the study results. In the case of a wind study, the effect of adding wind energy can be estimated by multiplying wind generation inputs by model market clearing prices (or market price inputs). The present value of the result generally is an upper limit on the reduction in model revenue requirements, as it does not account for additional opportunity costs from increased reserve requirements. Similarly, the difference between the model reduction in revenue requirements and the estimated revenue requirements should reflect the cost of withholding additional reserves, and other wind-related costs. Generally the result is a lesser reduction than the market price estimate by somewhere around 2–10% for most systems. Values outside or near the edges of such a range should be scrutinized carefully. Power systems with relatively large wind penetration rates and limited markets should tend to be near the upper end of the range, while systems with relatively lower levels of wind and access to large liquid markets should be closer to the lower or middle of this range.

A couple of operational aspects of the study results should also be examined. Most CEDMs have some form of reliability indicator such as loss of load probability (LOLP), loss of load expectation (LOLE), or energy not served (ENS). These indicators should be checked across studies to ensure especially that none of the resource strategies results in a substantial degradation of overall reliability. Holding too much or too little reserves out for wind could cause the model to encounter reliability issues. If this should occur, the analyst will need to ensure that the results are not due to modeling artifacts and are due to actual operational issues. The treatment of reserves may need to be adjusted to retain reliability metrics within acceptable bounds.

Shortfalls can also occur due to lack of ramping capability—that is, the ability to increase or decrease generation levels quickly enough to accommodate the combined effects of wind and load. It may be necessary to do a separate ad hoc analysis of ramping requirements to ensure sufficient capability exists. Another source of modeling error can occur where models assume too much flexibility (depth and ramp capability) in markets, effectively relying too much on markets to supply balancing services. This may be something of a judgment call depending on the liquidity and depth of available markets for a particular power system.

8.8 OVER-SPECIFICATION OF WIND COSTS

As wind power has come under intense scrutiny, analysts have tried to ensure that every potential cost caused by wind generation is properly accounted. On the one hand, this seems basically reasonable; however, there are similar costs borne by power systems that are not necessarily specifically called out for other resources or loads. Two examples of this are the unit-commitment costs associated with forecast error, and the regulating reserve requirement.

In the case of unit-commitment costs, similar costs could be ascribed to other resources, primarily those fueled by natural gas. In the USA, natural gas purchases are typically required a day in advance of delivery. Choosing the level of gas purchase involves estimating one or both of the expected next-day power demand or the next-day market prices. Incorrectly forecasting demand or market prices incurs an economic cost similar to that identified as unit-commitment costs for wind projects. However, there is little evidence that such costs are ever ascribed to natural gas-fueled power plants in their economic evaluations.

In a similar vein, regulating reserve costs are usually spread over the entirety of power system energy deliveries without respect to the individual contributions to the need for such reserves. For example, some large industrial loads are relatively constant over an hour or more, while others may vary wildly. Most power systems do not charge more for the greater loads with greater variability. In fact, the variability of wind on the timescale relevant to regulating reserves (seconds or minutes) is a minor component of the overall wind valuation analysis and some studies have chosen to omit explicitly analyzing them.

Most studies find that the significant factors in wind valuation are the base amount and timing of the energy generated, hour-ahead forecast uncertainty, intra-hour variability (of the order of 10 minutes to an hour), and the opportunity cost of additional reserve requirements. Some studies have found day-ahead forecast error leading to incorrect unit-commitment decisions one of the most significant factors, but this may be somewhat power system dependent. Regulating reserve requirements (reserves responsive on the order of seconds) have not been found to be a significant factor for the most part.

Finally, most valuation studies calculate the incremental reserves required to meet the additional difficulty balancing the system with significant levels of wind generation. However, there may already be standard ancillary service charges on wind generators that are not necessarily consumed by wind generation. In the USA, balancing areas normally require transmission customers to either contribute to contingency reserves—reserves required when sudden large outages occur—or purchase them from the transmission provider. However, wind generation is typically geographically dispersed enough that such outages are not credible events for wind projects. Contingency reserve requirements may be charged to wind generators as transmission customers but are not available to compensate for rapid loss of wind energy due to the rapid calming of wind. Reserves for those events are currently separately counted (in the USA) and charged.

8.9 SUMMARY

Economic analyses using complex chronological economic dispatch models (CEDMs) are commonly used in the valuation of competing resource choices. Studies employing CEDMs capture economic consequences stemming from the power system's response to wind generation that would not otherwise be easy to determine. Wind valuation studies are typically organized in pairs—often a status quo run over some study horizon, and a second study incorporating some level of added wind generation. Commonly available dispatch models may need modifications in order to capture costs associated with wind variability and forecast error. Multiple studies of this type may be run under different future scenarios (especially fuel costs) to examine potential revenue uncertainty (risk) reductions associated with wind generation. Some costs not captured by CEDMs may need to be added separately into the valuation study. Due to the complexity of power system models for large power systems, it is important to validate model inputs, logic, and results. It is important to validate against estimated wind generation value and costs. Some of the costs that have been associated with wind are sometimes overlooked for other resources and loads, and calls to question the importance and fairness of attributing such costs directly to wind generation.

Chapter | nine

Wind Integration Costs

Problems worthy of attack prove their worth by fighting back.

Paul Erdös, mathematician, 1913–1996

Wind integration costs are somewhat loosely defined as those costs associated with integrating wind generation into power systems that are specific to wind, or that are minimal with respect to other generating resources and that do not include the value of the energy itself. Computing these costs is not a necessary component of valuing wind generation as should be clear by the discussion in the previous chapter. However, there are several reasons that identifying wind-specific costs may be desired or necessary. An often-cited statistic used to compare competing resource technologies is the levelized cost of the energy produced. Those costs traditionally include capital, fuel, operation and maintenance (O&M), and financing costs. There may be a desire to include integration costs in the comparison for wind generation. Balancing area operators may seek wind integration costs in order to assess wind-specific tariffs to recover ancillary service costs that would not otherwise be captured through the existing rate structure. Finally, wind integration cost studies can be quite complex, and it may be useful to rely on pre-computed wind integration costs or specific cost components to simplify treatment of wind in the usual resource planning studies.

The biggest contributor to wind-specific integration costs is generally found to be the cost of maintaining and operating higher levels of reserve generation. Some studies have found costs due to unit-commitment errors associated with longer-term (many hours to a few days in advance) forecast accuracy to be a significant, or even a dominant, cost factor. Exactly what constitutes a wind integration cost may depend partly on how the costs are to be used. For example, if the result of the analysis is to be a cost adder to the results of a chronological economic dispatch model (CEDM), then only those costs not specifically captured by the model should be used.

Valuing Wind Generation on Integrated Power Systems. DOI: 10.1016/B978-0-8155-2047-4.10009-2

Care should also be taken in not over-identifying integration costs for wind that are not separately calculated for other resources or for comparable issues with respect to load. For example, while some industrial manufacturing loads may be quite predictable, others can be relatively unpredictable (e.g. electric arc furnaces or irrigation pumps) without distinguishing rates established to differentiate among them. The analyst must work with policymakers to be certain that treating wind in this way is appropriate and not unduly discriminatory.

9.1 WIND INTEGRATION COST STUDY DESIGN

Wind integration cost studies have much in common with the valuation techniques discussed in Chapter 8. The difference is that there are usually just two comparison model runs—one including a pattern of wind generation as it might actually occur over the study horizon, and a second study with an equivalent amount of energy but with some or all of the variability and uncertainty removed. The purpose of having the second ('flat') study is to isolate operating and dispatch costs related to the variability and uncertainty of wind generation. A portion of the wind integration costs is the cost difference between the energy equivalent study and the wind pattern study.

9.1.1 Design for CEDM-based studies

A simplified study design using a chronological economic dispatch model (CEDM) is shown in Figure 9.1. As discussed in Chapter 8, the individual CEDM run blocks depicted in Figure 9.1 may in fact entail multiple runs to fix unit-commitment decisions depending on the sophistication of the model. The left side of the figure represents determining the present value costs assuming a pattern of wind generation as input to the model. Incremental reserve requirements consistent with the methods described in Chapter 6 are used as input to the model as well. The right side of Figure 9.1 shows a second model run in which the same amount of energy is assumed as in the first study, but without much of the variability and uncertainty inherent in the first study.

Studies differ on how the energy equivalent is represented in the second study. Some studies adopt a constant level of generation or contract purchase whose annual energy is equivalent to the studied wind resource for the comparison model run. However, representing the energy equivalent as a single value through the year can introduce its own bias into the result. This is illustrated in Figure 9.2, where the variations in market prices over a year happen to coincide relatively closely with the pattern of wind generation being examined. In other words, wind generation tends to be higher in months where prices happen to be high and lower in months where prices are lower. On the other hand, the flat block of equivalent energy will be more reflective of the average market price over the year. The difference between the two generation patterns is not indicative of an integration cost for wind—merely the relative value of the timing of the generation that would normally be captured in a valuation study. The wind data reflected in Figure 9.2 happen to be more valuable than the flat delivery pattern, which would tend to incorrectly offset the other wind integration costs—in other words, bias the integration costs downward. The bias can go either way depending on the pattern of the wind generation being evaluated, and in either case the bias needs to be removed in some way.

A similar problem can be introduced due to the diurnal shape of wind generation, as market prices tend to be higher during the day than at night. Comparing a wind project that generates at greater rates on average at night than during the day will have a lower value than a constant rate of

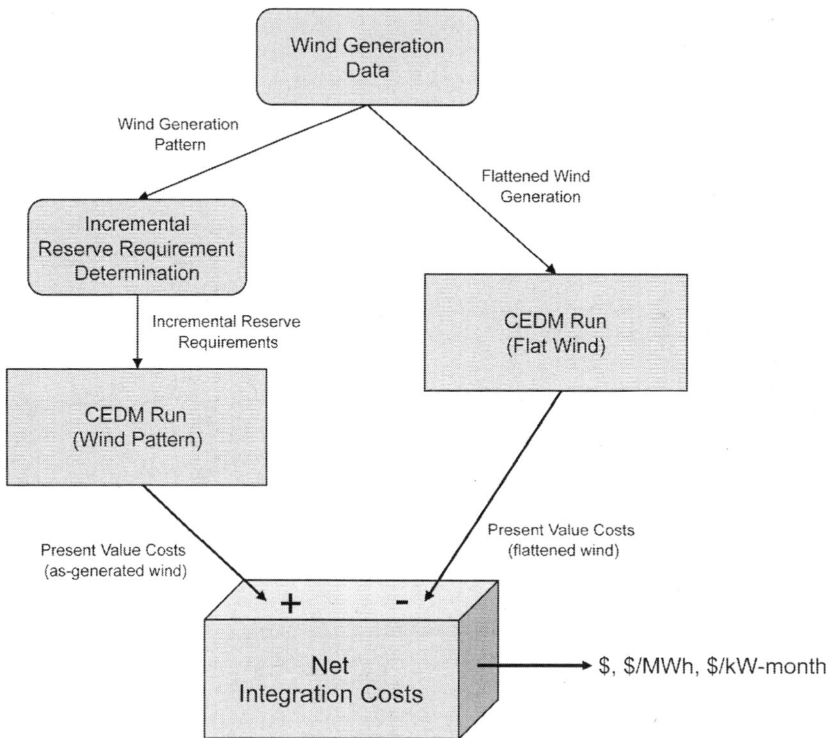

FIGURE 9.1 Steps in developing wind integration costs. Two CEDM runs are performed, each with similar amounts of energy. The left side of the diagram depicts a study run with the wind generation pattern directly. On the right is a second study that is run with the wind energy in flat blocks as discussed in the text.

delivery of the equivalent energy. If not otherwise taken into account, the wind integration cost could inadvertently be overstated. It is important to remove this effect as the value of the wind energy, including the timing of its delivery, should be separately captured in the valuation study outside an integration cost analysis.

One way to address this problem is to compare the wind pattern to an equivalent amount of energy that changes both by month and by time of day—for example, break the wind pattern into daytime and night-time blocks of energy for each month. There are potential problems with this approach as well, in that it may not remove all of the bias (though most of it should be accounted for), and introduces an amount of variability that begins to approach that of the wind generation pattern itself. For now the best options appear to be explicitly calculating and removing the bias from the result, or to accept some amount of variability in the comparison model runs introduced by using monthly or weekly daytime and night-time energy blocks. A third possibility is that if the wind pattern does not happen to have strong diurnal or seasonal patterns, the flat block may be appropriate. In general, however, the flat block will introduce a bias that needs to be removed.

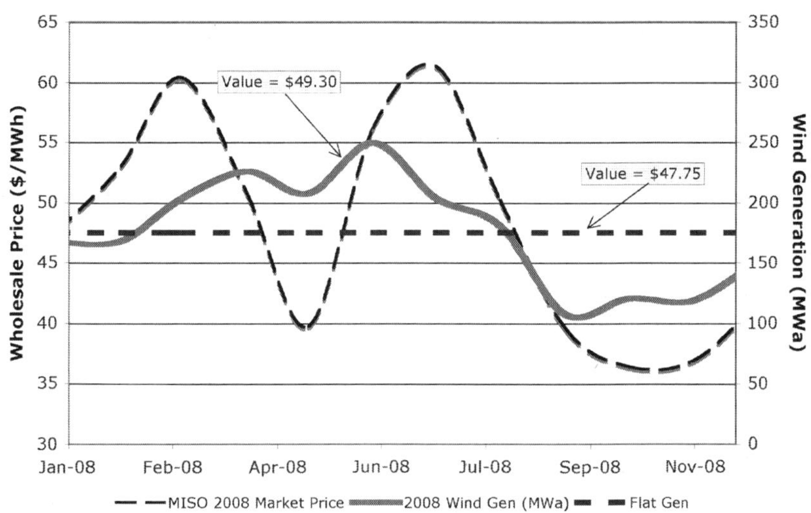

FIGURE 9.2 The timing of wind gen2eration and market prices through the year can artificially introduce a phantom integration cost. In this case, the average monthly wind generation levels are actually more valuable than a flat delivery by $1.55 MWh and would potentially reduce the otherwise computed integration cost by that much. In other cases, wind generation can be less valuable than the flat curve, artificially increasing the computed integration cost.

9.1.2 Non-CEDM study design

Not all wind integration cost analyses are performed using the relatively complex power system operation representations comprehended in CEDMs. Figure 9.3 shows a study design for such an analysis. In this design the costs are divided into components that include:

- Increased day(s)-ahead unit-commitment or market purchase costs
- Increased hour-ahead unit-commitment or market purchase costs
- Cost of holding and operating additional reserves.

As Figure 9.3 suggests, calculating these costs begins with data sets representing both wind generation and load through the study period, as well as wind and load forecasts. Developing these data sets is described in Chapters 5 and 7. Once the wind/load generation and forecast data are available, day-ahead and hour-ahead balancing costs can be determined. It is necessary to minimize the amount of expensive reserves needed through adjustments made prior to the operating period. Adjustments include a combination of unit-commitment decisions and market transactions in the day-ahead and hour-ahead timeframes to minimize the amount of reserves ultimately needed.

The largest wind integration cost is usually associated with holding and operating additional reserve-generating units. Calculating the cost of holding and operating reserve generation begins with determining the quantity of additional reserves needed. Chapter 6 details how this is done, taking account of similar reserves needed to accommodate the variability and uncertainty of system demand. These are statistical computations and not dependent on running CEDMs.

Once the quantity of additional reserves is determined, some value must be assigned to them. As described in Chapter 8, one way this can be done is with a CEDM study in which the model is instructed to withhold the incremental reserves. The resulting cost increases calculated by the model represent opportunity costs from withholding generating units from market sales (or reducing market purchases) to ensure adequate system reliability. There may also be costs accruing from the need to hold more generating units in a ready ('spinning') condition, and reductions in overall efficiency that can all be calculated by a reasonably sophisticated CEDM.

Conversely, the value of reserves may be available from historical market information. In the extreme, proxy resources can be developed for the express purpose of providing the reserves. Such a computation is illustrated in the example below. Large power systems find it more economic to spread the reserve burden among a number of power plants, enabling each individual generator to operate near its maximum output level and minimize the overall reduction in efficiency. Concentrating the reserve requirement on one or a few generators forces them to

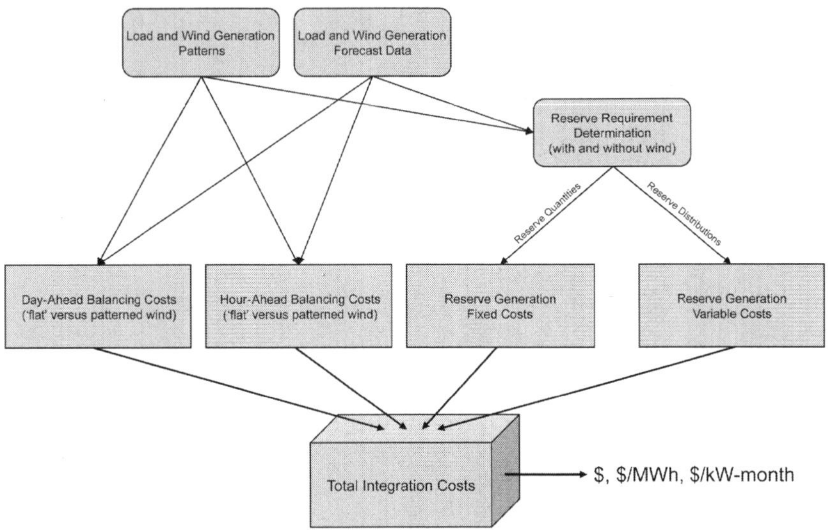

FIGURE 9.3 A study design that does not depend on chronological economic dispatch models (CEDMs). Day-ahead and hour-ahead unit-commitment (or market transaction) costs are determined on the left half of the diagram, while the fixed and operating costs of reserves are determined on the right half.

operate near the middle of their capability, where the efficiencies are relatively low.[1]

9.2 SIMPLIFIED NON-CEDM WIND INTEGRATION COST EXAMPLE

It may be useful to step through the process of developing wind integration costs using a simplified data set. The data used in this example will be confined to just two consecutive days, to make it easier to visualize the data and analyses. To further simplify, we will take the rather extreme example of a system with access to markets that trade in day-ahead and hour-ahead time periods, but for which all remaining balancing services are provided by dedicated generating units. It is further assumed that the incremental balancing reserve requirement will be met by purchasing additional dedicated reserve-generating capability expressly to provide the computed incremental reserve requirement. The wind resource in this example has a nameplate generating capability of

[1] Many thermal units are not designed to operate below roughly the 50% capability level and the reserves must be spread out, or else held on generators specifically designed to be able to operate at low minimum levels.

800 MW, and an annual energy production of 2.6 million MWh for a 27% capacity factor.

9.2.1 Calculating increased reserve requirement

Two days of sample wind and load data are shown in Figure 9.4, along with the hour-ahead schedules. Reserve generation is set aside to accommodate (most of) the net of the differences between observed and scheduled wind and load. For the purposes of this example, reserves are sized to accommodate 95% of these differences to meet reliability requirements. The net differences between observed and hourly schedules for wind and load are shown in Figure 9.5.

The incremental reserve requirement is determined by finding the 95% reserve levels needed for the combination of wind and load and subtracting the amount that would be needed for load alone. In this example, the requirement for combined wind and load is 214 MW of incremental generation capability and 253 MW of decremental generating capability, as shown in Figure 9.5. The similar requirements for load without taking wind into account are 186 and 235 MW of incremental and decremental reserves respectively. The reserve requirements introduced by the wind can therefore be taken as 28 MW of incremental capability (214 - 186 MW) and 18 MW of increased decremental capability (253 - 235 MW).

9.2.2 Incremental fixed costs

Assume that the methods described above, and detailed more fully in Chapter 6, reveal a need to add 25 MW of additional reserve requirements. That represents the need to increase and decrease generation by 25 MW over the expected operating period generation level. The utility plans to meet this requirement by acquiring three 30-MW simple cycle generating units with a levelized fixed cost of $150/kW-year, or a total annual cost of $13.5 million (3 × 30,000 kW × $150/kW-year). One of the units has the sole purpose of providing capability during outages of either of the other two generating units. The units are sized at 30 MW instead of 25 MW to accommodate minimum allowable operating levels of the units.

Translating the annual cost of $13.5 million into a dollar-per-megawatt hour of wind generation necessitates dividing by the annual wind energy production. Dividing by the assumed annual wind energy production of 2.6 million MWh results in fixed wind integration costs in this example of $5.19/MWh. This forms an upper limit[2] on wind integration fixed costs, assuming that all incremental reserve requirements

[2] No additional value is accorded to the gas plant due to the optionality provided when not needed for wind—for example, either because the wind is calm for some period or due to the seasonal nature of the wind requirement.

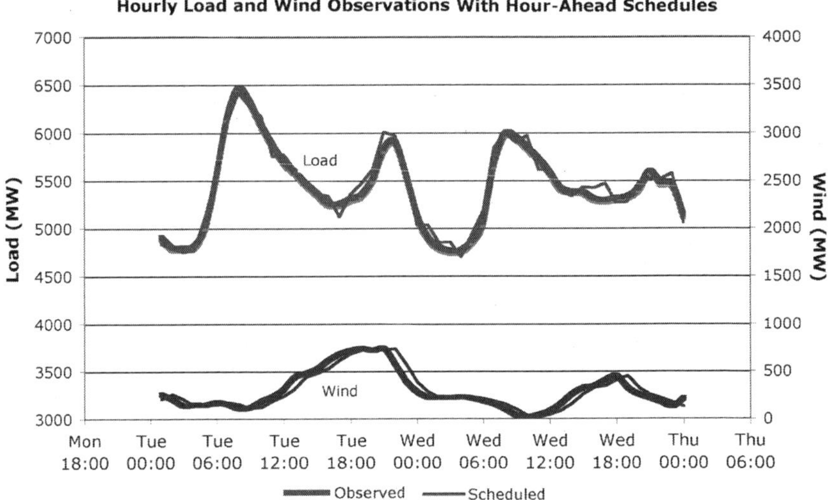

FIGURE 9.4 Wind and load data used for determining example wind integration costs. Heavier lines indicate observed values and lighter lines represent hour-ahead schedules based on forecasts of the observed quantities.

come from new generating units.[3] The cost can also be put in terms of nameplate wind generating capacity. In this example, the wind capacity is 800 MW (800,000 kW) and the $13.5 million levelized fixed costs can alternatively be characterized as $1.41/kW-month ($13,500,000/800,000 kW/12 months/year).

9.2.3 Incremental fuel costs

In addition to the fixed costs of incremental reserve capability, the increased operations on balancing generators will increase overall fuel consumption primarily due to lower average efficiency of all generating units providing balancing services. Figure 9.6 illustrates an aggregate heat rate curve for generating units providing balancing reserves for this example. Two important features of the heat rate curve are: the heat rate is lower (most efficient)near the maximum output level, and the rate of decline is not linear throughout the operating levels. Although the data shown in Figure 9.6 are merely an example, declining heat rates and nonlinearity are general features of heat rate curves.

Fuel consumed to provide balancing services grows as the balancing requirements increase. This effect is due to the nonlinearity of the heat rate curve. Since the heat rate at lower generating levels is disproportionately

---

[3] In the alternative, additional reserve could be held on existing units with resulting energy and capacity being replaced with a contract purchase or incremental baseload generation.

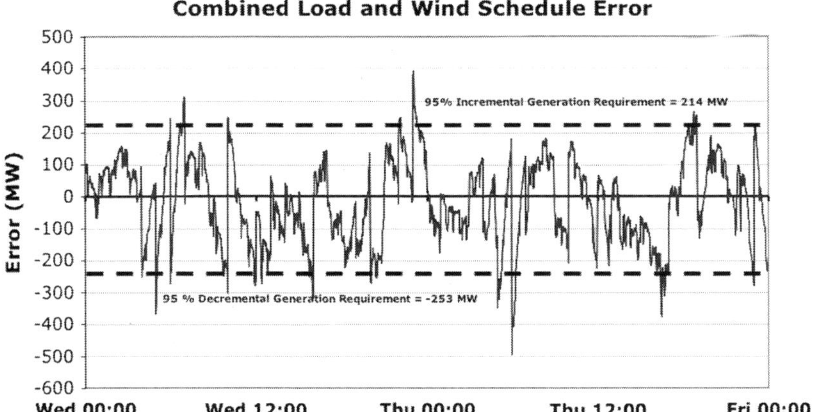

FIGURE 9.5 The net differences between observed and scheduled load net of wind generation. Reserve requirements are determined by finding the upper and lower limits containing 95% of all values. In practice, this would be set on a larger data set than the two days used for illustration purposes here.

higher than the reduction in heat rate at the upper range, the overall effect is to further reduce the overall efficiency. For the 2-day sample data represented in Figures 9.5 and 9.6, the overall fuel consumption increased 4 BTU/kWh as the average heat rate increased from 10,005 to 10,009 BTU/kWh, as illustrated in Figure 9.6.

Average heat rate is computed by taking a weighted average based on the expected distribution of the balancing generator operations. Figure 9.7 illustrates the distributions of balancing generator levels based on the data presented in Figure 9.5. The percentages in Figure 9.7 for each operating level are multiplied by the heat rate for those levels illustrated in Figure 9.6. The result is a relatively small increase in overall fuel consumption. In this example, there is an increase of 4 BTU/kWh (4000 BTU/MWh) to be applied to the average operating level (350 MW).

To calculate a component of the wind integration costs due to the reduction in overall balancing generator efficiency, the increased fuel consumed is translated to an annualized cost and spread over the wind nameplate capacity or energy production. Since the average generation from the balancing generators in this example is 350 MW, the increased fuel consumption is 12,264 million BTU/year (4000 BTU/kWh × 350 MW × 8760 hours/year). If the fuel cost is $10/million BTU, this equates to $122,640 per year increased fuel costs. Spreading those costs over 800 MW of nameplate wind capacity results in a wind integration cost contribution of about $0.013/kW-month. This is a small fraction of the fixed cost component ($1.41/kW-month). Alternatively, expressing the costs on a wind energy production basis $0.047/MWh

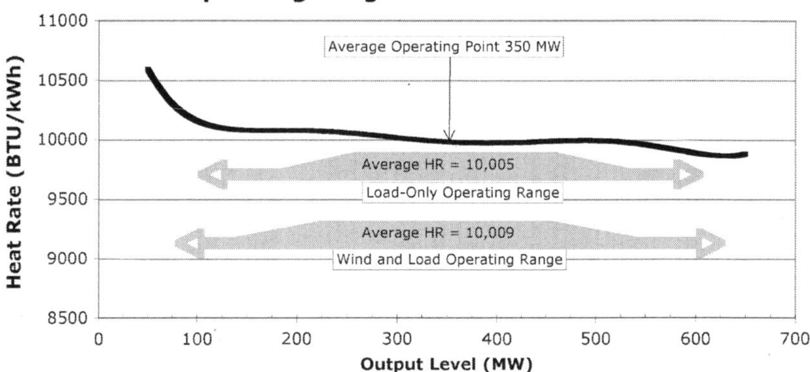

FIGURE 9.6 The effect on overall heat rate of balancing generation when the balancing requirement increases from 250 (plus and minus) to 275 MW. For the load-only case, 500 MW of reserve unit capability operates between 100 and 600 MW, with average generation at 350 MW. In the wind and load case, the reserve requirement has risen to 275 MW and the operating range extends from 75 to 625 MW, reflecting the 25 MW increase in reserve capability added to the system. The increased range of operation reduces the overall efficiency of the reserve generators.

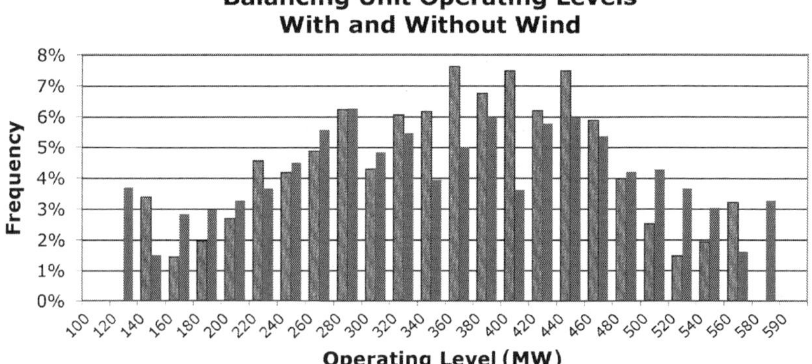

FIGURE 9.7 The distribution of 1-minute balancing requirements, limited to the available reserve generation as determined by the 97.5 percentile levels of the 1-minute data in Figure 9.5. The relative frequency of operating levels are used in combination with the power system equivalent heat rate illustrated in Figure 9.7 to determine the weighted average heat rates.

($122,640/2,600,000 MWh/year)—again, a small percentage of the fixed costs of $5.19/MWh.

9.2.4 Market transaction costs

The histogram of reserve unit operating levels illustrated in Figure 9.7 is the remaining variability after market purchases or sales conducted to help balance the system. The relatively high cost of balancing reserves motivates minimizing that need by balancing the system to the extent possible in wholesale markets (or with owned units) ahead of the operating period. Further, it is usually less expensive to transact a day in advance of the delivery period, compared to transacting in the marketplace for the next hour. Power system forecasters have only a general idea of what to expect for system load and wind generation many weeks or months in advance. Nevertheless, planning to balance the system is undertaken to reduce the difference between the expected generating capability and demand for the operating periods weeks and months in advance in order to reduce the near-term (days or hours) adjustments that may be necessary.

For wind generation, planners may assume a long-term average for the expected generation level. Some systems may assume zero wind output from the wind generators and rely on nearer-term markets for balancing—preferring to be in a long (surplus) position rather than a short (deficit) position as the operating period approaches. Figure 9.8 illustrates the market transactions necessary to balance the system based on the day-ahead wind forecast.

A similar situation exists for system load balancing. Figure 9.9 illustrates the day-ahead balancing requirements for load. Note that on at least some hours, the net market transaction is reduced because the wind and load needs cancel one another. This is an important effect that needs to be taken into account in any analysis. In this particular example, the day-ahead forecast suggested system demand systematically lower than the longer-range prediction.

Balancing the individual requirements independently of one another would require transacting on 7600 MWh of purchases and sales for wind and 22,400 MWh for load. Figure 9.10 shows the net transactions required for the combination of wind and load amount to 28,400 MWh. Note that the combined transactions are lower than the sum of the individual transaction (30,000 MWh) because the needs cancel at least partially on some hours. It is certainly possible that the needs would only add, but in general the net requirements will be lower than the sum of the individual requirements because it is unlikely that load and wind adjustments will be additive on all hours.

Costs attributable to the increased transactions are considered a component of wind integration costs in some studies. A similar procedure is employed to move from the day-ahead transactions to

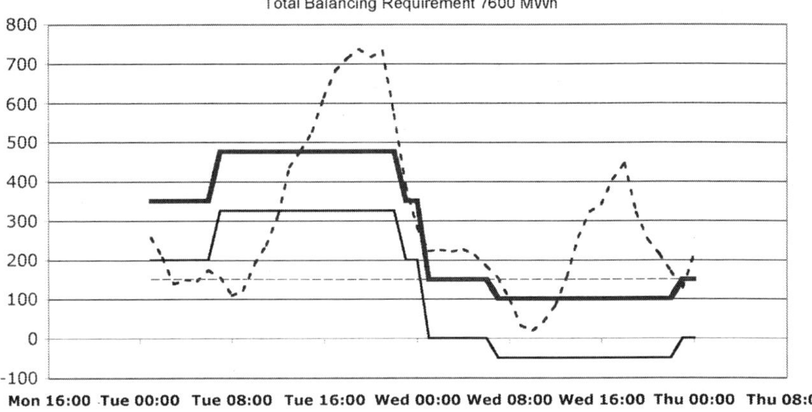

FIGURE 9.8 The market transactions undertaken in the day-ahead timeframe as wind energy forecasts give better information about the expected wind generation. The seasonal long-term average wind generation is represented by the straight dashed line near 150 MW. The thick solid line shows the day-ahead wind forecast, and the thinner solid line shows the purchases (positive values) and sales (negative values) needed to move from the seasonal expectation to the day-ahead forecast prior to accounting for the similar requirement for load.

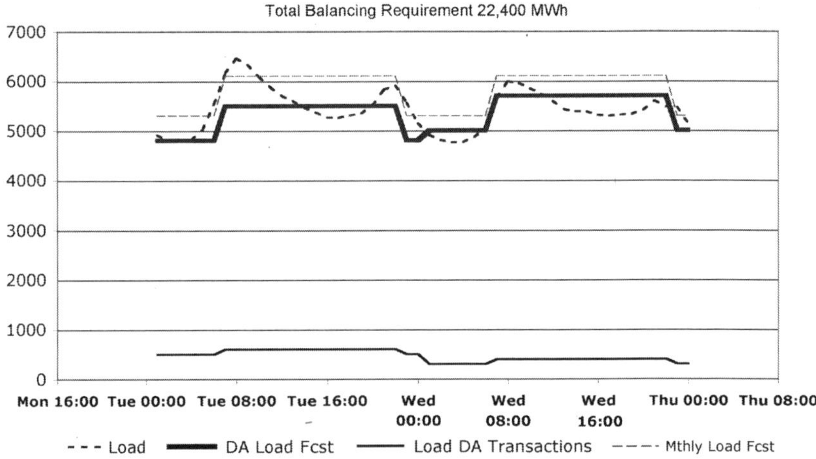

FIGURE 9.9 The adjustments to be made to bring the system balance closer based on day-ahead load forecast. Total balancing requirement reflects the sum of the MWh adjustments between the longer-range forecast and the day-ahead forecast.

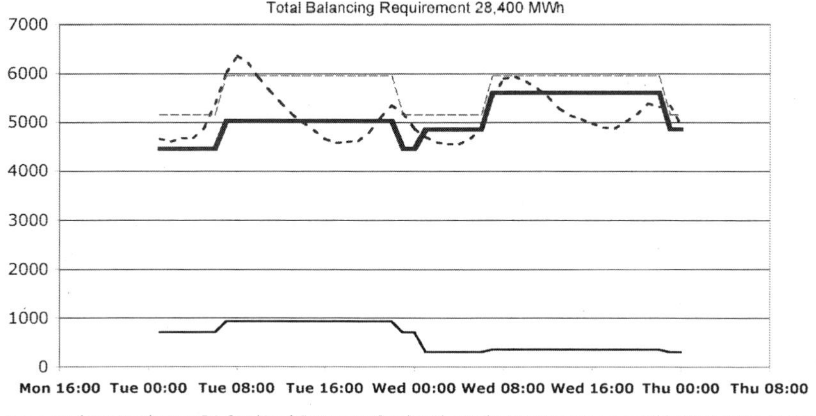

FIGURE 9.10 The transaction requirements after netting the individual wind and load balancing needs shown in Figures 9.8 and 9.9. Combined balancing transactions for wind net of load are less than the sum of those that would individually be required for wind and load separately.

hour-ahead transactions based on nearer-term forecasts for load and wind as they become available. For the sample data used here the net balancing transactions in the hour-ahead time frame were 10,830 MWh and those for load alone were 22,400 MWh. Transactions for load net of wind were 28,400 MW signifying an increase of 6000 MWh for this 2-day period, or 3000 MWh average for each day, or 1,095,000 MWh per year (3000 MWh/day × 365 days/year) if the two days examined are representative.[4]

A transaction cost must be ascribed to transacting in these markets to determine the contribution to wind integration cost. For the purposes of this example, assume that each megawatt-hour of energy transacted incurs a \$0.50/MWh transaction cost for day-ahead transactions. If the two days examined here are representative of the increased transactions over a year, the total increased day-ahead transaction costs would be \$0.5475 million per year. Expressing this value in terms of wind energy produced results in a contribution of \$0.21/MWh (\$0.5475 million per year/2.6 million MWh per year) due to increased day-ahead purchases. On a nameplate capacity basis this can be expressed as \$0.57/kW-month (\$547,500/800,000 kW per 12 months/year).

[4] This example is not meant to suggest that 2 days of data are sufficient to determine the wind integration cost. In general, at least a full year of data should be examined if possible. The purpose of limiting the data set to 2 days in this example is merely to allow the reader an easier time visualizing the data and data analysis.

The market value of the power transacted does not enter into this analysis as it is assumed that the load and wind forecasts are unbiased (i.e. the average error is near zero), implying that the net energy transacted is essentially zero. There will of course be months or years where the energy transacted has a net positive or negative value, but over the long term the average should tend toward zero and is not further considered here.

An exactly analogous procedure is used to develop incremental hour-ahead transaction costs—evaluating the increased transactions to move from the day-ahead balancing to the hour-ahead forecasts. Using the two days of test data presented here, the wind-balancing requirement reduced the overall transactions so the net effect would be to reduce transaction costs, but this will not be generally true. There will likely be some increase due to the wind-balancing requirements, but that increase may be small if the day-ahead wind schedules are reasonably accurate or the overall wind on the system is relatively small.

9.2.5 Summary of costs

Table 9.1 shows the various costs computed for this example. Note that the levelized costs quoted implicitly assume a constant level of wind on the system through time. More likely, the amount of wind on a system will be assumed to increase through time, and the costs would change. Integration costs are sensitive to the relative level of variability and uncertainty between wind and load, as should be clear in the example everywhere load- and wind-balancing requirements were netted. In other words, integration costs even for a given set of wind resources can significantly differ depending on the power system into which they interconnect. In this example, 800 MW of wind were interconnected to a system of roughly 7000 MW of peak load, or in 2.5% of energy generation coming from wind

TABLE 9.1 Integration Costs from the Worked Example

	Example Wind Integration Costs	
	$/MWh	$/kW-month
Incremental reserve generation fixed costs	$ 5.19	$ 1.41
Incremental reserve generation fuel costs	$ 0.05	$ 0.01
Day-ahead balancing costs	$ 0.21	$ 0.06
Hour-ahead balancing costs	$ —	$ —
Total	**$ 5.45**	**$ 1.48**

generation. On a significantly smaller system, 800 MW of wind would be associated with higher costs.

9.3 COST ALLOCATION

Wind integration cost studies typically focus on the incremental costs when wind is added to a power system. Wind integration costs as described above are not independent of the characteristics of the existing load on the system and ascribing the incremental requirements all to wind on the system is mainly an artifact of wind additions being a relatively new phenomenon. If wind had been added to power systems through the years as load increased, calculating wind integration costs as incremental to load would not be obvious or natural. There would be no reason to equate the incremental with the wind resources.

Various means of dividing up the total reserve requirement can be devised. In the example of Figure 9.2, the incremental reserve requirement was seen to be approximately 25 MW out of a total requirement of approximately 250 MW. It is possible to conclude, as most studies do, that the wind requirement is therefore 10% of the total. However, a simple algorithm that assigns a proportionate responsibility of the total reserve requirement on the basis of the individual contributions suggests that, on average, 37% of the reserve requirement for any particular time interval is attributable to wind and 63% to load.[5] This is because much of the time reserves needed for load in the absence of wind are accessed by wind, and vice versa.

There is no objectively correct way to allocate the reserve requirements that are shared between load and wind. It may be necessary to adopt some allocation methodology for rate-making purposes. It is unclear what the fairest allocation should be. Wind, as the incremental resource causing an incremental need, may be (and usually is) ascribed the incremental cost of holding reserves. It might equally be argued that the allocation should be made on the basis of actual use of reserves, which would tend to allocate a higher percentage.

Whatever allocation scheme is adopted should be based on achieving a specific, identifiable objective. One such objective would be to reflect the incremental cost or benefit from a load or generator back to that load or generator. Designing tariffs and rate determinants that provide proportionate economic signals and that can be practically implemented presents its own challenge. Nevertheless, balancing area authorities are generally interested in providing appropriate incentives reflective of relative

[5] In time periods where the errors were in opposite directions (i.e. canceling one another), 100% was allocated to the larger absolute schedule error and 0% to the smaller one.

integration costs. The following section provides guidance on how reserve requirements change under the addition (or subtraction) of a load or generator to an existing system.

9.4 INCREMENTAL RESERVE REQUIREMENT BEHAVIOR

Wind integration costs are tied directly to the amount of reserve requirements needed to maintain system reliability criteria as wind generation is added to the system. While Chapter 6 details how reserve requirements are directly calculated from wind and load data, it may be useful to understand more generally how reserve requirements change for rate-making purposes.[6]

As may be clear by now, the integration cost of a particular wind project is not an inherent characteristic of the wind project itself, but depends also on the characteristics of the power system into which it connects. Relevant characteristics relate both to the existing system loads (and their variability) and the other wind projects that exist on the system. This implies that the incremental reserve requirement for a specific project is different depending on the order in which it is added. This section explores some of the relevant properties of incremental reserve requirements arising from added projects. Specifically, a method is shown for an analytical approach that determines the relative importance of generator size, correlation with the existing system schedule errors, and overall variability.

9.4.1 Importance of standard deviation

Tracking how reserve requirements change entails examining the behavior of standard deviations. Figure 9.5 illustrates the genesis of reserve requirements from schedule errors. Reserve levels are selected such that some percentage (e.g. 95%) of all schedule errors is contained within the selected levels. The data represented in Figure 9.5 show two days of sample 1-minute schedule error data. The standard deviation of these numbers is 129 MW, and the reserve requirements defined by the 95% confidence level are +214 and −253 MW, or alternatively

[6] Wind integration charges are not yet common in North America or in Europe. There are pros and cons to instituting such charges, and fairness arguments can apply to both sides. For example, no 'integration costs' are accorded large nuclear plants in proportion to their impact to contingency requirements, and no unit-commitment charges are levied on natural gas units when day-ahead price forecast errors result in operating losses. The societal interests of adopting renewable generation may be deemed to outweigh the potential perceived benefits accruing from assessing a charge to better reflect cost causation or to incentivize less volatile or more easily forecasted wind sites.

represented as +1.6 and −1.8 standard deviations around a mean schedule error of −16.5 MW.

The standard deviation is one defining characteristic of any data set. Standard deviation is a general measure of the dispersion of numbers about their mean. Another statistic used to characterize a distribution of numbers is to quantify the quantiles as some distance from the mean in terms of the standard deviation. For example, the values represented in Figure 9.5 have 95% of all values lying between +1.6 and −1.8 standard deviations about the mean. These values are termed *z*-statistics, and are especially important in normal probability distributions because there is a direct association between a *z*-statistic and a confidence interval. For example, if a set of numbers is normally distributed, 95% of the values are expected to be found within ±1.96 standard deviations (i.e. $z = \pm 1.96$) of the mean—irrespective of the particular distribution's mean and standard deviation.

This implies that the reserve requirement is proportional to the standard deviation for normally distributed schedule errors. In practice, the errors are not normally distributed, and a 95% confidence interval may be represented by a *z*-statistic different than ±1.96 (as in our example, which is also not symmetrical). However, it is reasonable to expect that for small changes in a distribution, the reserve requirement will be approximately proportional to the standard deviation.

For example, if the reserve requirement prior to an addition is 25 MW, and an addition increases the standard deviation by 10%, then the effect would be to increase the reserve by 10% as well, to 27.5% in this case.[7] While this is identically true for normally distributed numbers, it may also be a reasonable estimate for small changes in distributions that are not normal. Understanding how the standard deviation changes when one set of schedule errors is joined by another set helps understand how reserve requirements change as well.

9.4.2 Summing distributions

Schedule errors such as those in Figure 9.5 can be summed with a set of errors that represent the addition of another wind project (or load increment) to form a new distribution, with a different reserve requirement and different standard deviation. The new errors will have their own distribution, and the numbers may or may not be correlated with the existing numbers. Correlation is a statistical computation that expresses the degree that two sets of paired numbers tend to move together.

The standard deviation of the new set of errors, constructed by summing the errors of the original system with those of the addition, is

[7] This assumes the schedule errors have zero mean.

a function of standard deviations of the individual series, and of the correlation between them:

$$S_T = \sqrt{S_o^2 + s^2 + 2rS_o s} \tag{9.1}$$

where S_T is the standard deviation of the combined schedule errors, S_o is the standard deviation of the original system, r is the correlation between the original errors and the added errors, and s is the standard deviation of the added schedule errors.

Letting the standard deviation of the added series (s) be represented as some fraction (f) of the original series standard deviation ($f = s/S_o$), the above equation can be rewritten as:

$$S_T = S_o\sqrt{1 + f(f + 2r)}$$

The change in standard deviation is then:

$$S_T - S_o = S_o\sqrt{1 + f(f + 2r)} - S_o = S_o\left[\sqrt{1 + f(f + 2r)} - 1\right]$$

The change in fractional standard deviation can be written as:

$$\frac{S_T - S_o}{S_o} = \frac{S_T}{S_o} - 1 = \sqrt{1 + f(f + 2r)} - 1 \tag{9.2}$$

This equation describes the relative effects of correlation and project size on the standard deviation of the combined errors. Figures 9.11 and

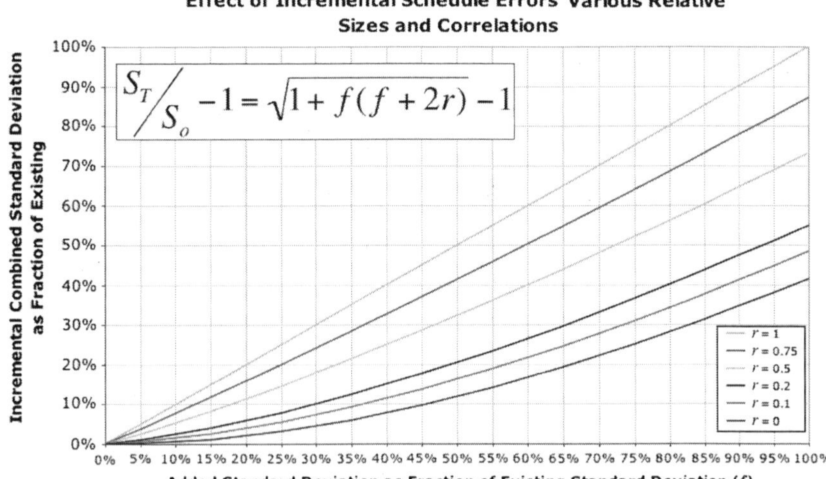

FIGURE 9.11 The increase in combined standard deviation as a function of the standard deviation of the added schedule errors. Both axes are in terms of a fraction of the standard deviation of the system prior to the added schedule errors.

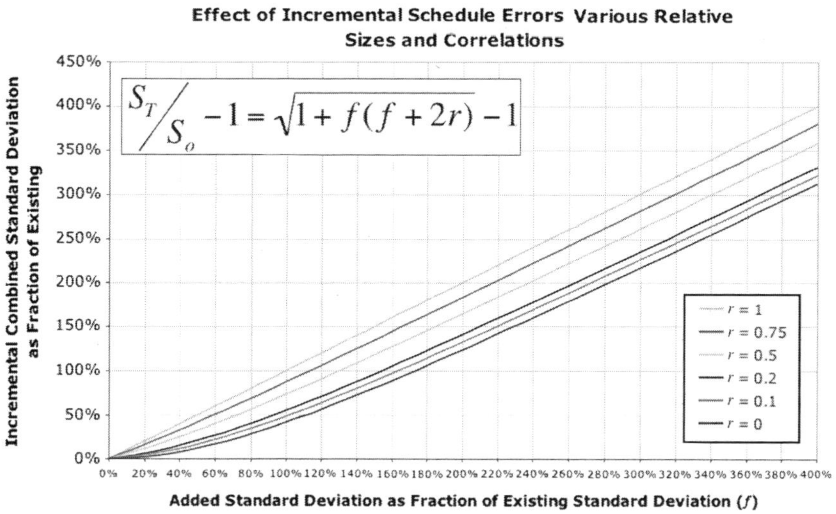

Effect of Incremental Schedule Errors Various Relative Sizes and Correlations

$$\frac{S_T}{S_o} - 1 = \sqrt{1 + f(f + 2r)} - 1$$

r = 1
r = 0.75
r = 0.5
r = 0.2
r = 0.1
r = 0

Incremental Combined Standard Deviation as Fraction of Existing

Added Standard Deviation as Fraction of Existing Standard Deviation (*f*)

FIGURE 9.12 The same as Figure 9.11 except on a larger scale.

9.12 graph this relationship for a range of correlations and addition sizes.

Figure 9.11 shows that when the correlation is identically 1.0 (top line), and the standard deviation of the added schedule errors is the same as the original distribution (i.e. f = 100%, r = 1.0), then the incremental standard deviation (y-axis value) is also 100%. In other words, the standard deviation of the combined system is double that of the original system. On the other hand, if the correlation were zero (f = 100%, r = 0), Figure 9.11 shows that the standard deviation of the combined system is 41% higher than the original system standard deviation.

9.4.3 Effect of project size: Examples

Consider a system with a current reserve requirement of 150 MW, where the standard deviation of schedule errors is 75 MW (i.e. z-statistic = 2.0 for acceptable reliability). Equation (9.2) can be used to estimate the increased reserve requirement (assuming normality) due to the addition of either of two wind projects: A and B. Project A is 100 MW with a schedule error standard deviation of 30 MW, and project B is 10 MW with standard deviation 3 MW. Schedule errors for both projects have a correlation with the existing schedule errors of 0.20 (i.e. r = 0.2).

The standard deviation of project A (30 MW) is 0.4 times the original system standard deviation (75 MW). Setting f = 0.4 and r = 0.2 into equation (9.2), or by inspection of Figure 9.11, the fractional change in the system standard deviation is 14.9%. The megawatt increase in reserve requirement is therefore 14.9% of 150 MW, or 22.3 MW. In terms of the nameplate wind capacity, this suggests that the incremental reserve requirement is 22.3% of nameplate capacity.

The same procedure can be undertaken for project B. In this case variable f in equation (9.2) becomes 0.04 (3 MW/75 MW), and r remains 0.2. Inserting these values into equation (9.2), the fractional incremental reserve requirement becomes 0.876%. This represents (150 MW × 0.00876) 1.3 MW of increased reserve requirement. Expressed in terms of nameplate capacity, the incremental reserve requirement is 13% of nameplate capacity in this case.

9.4.4 Effect of correlation: Examples

Consider two projects A′ and B′ that are similar to projects A and B in all respects except that the correlation between the schedule errors for A′ and B′ is 0.1 instead of 0.2 as before. The incremental reserve requirement for A′ goes from 14.9% to 11.4% of the original reserve requirement (150 MW). This amounts to 17 MW, or 17.0% of nameplate, down from 22.3% computed for a correlation of 0.2.

Similarly, project B′ moves from a fractional reserve requirement increase of 0.876% to 0.479% of 150 MW, taking the increased reserve requirement of 1.3 MW to 0.7 MW—a reduction that is roughly proportional to the change in correlation.

9.4.5 Small increment approximation

It should be noted that equation (9.2) can be approximated closely for small (<<1) values of f:

$$\frac{S_T}{S_o} - 1 = \frac{f^2}{2} + rf, \qquad f << 1 \tag{9.3}$$

Applying this approximation to projects B and B′ gives respectively 0.88% and 0.48% compared with the more precise calculations of 0.876% and 0.479% calculated above using equation (9.2) directly. The simplification in equation (9.3) may be useful for developing simplified rate determinants that capture some of the relatively complex interplay between project size and diversity (i.e. level of correlation).

Equation (9.3) can be used to show the approximate incremental reserve requirement as a fraction of installed capacity for relatively small projects. Let α represent the ratio of the incremental project's schedule error standard deviation to the original system standard deviation:

$$\alpha = s/P$$

where s is the incremental project's standard deviation of schedule errors and P is its nameplate generating capability.

The incremental reserve requirement as a fraction of nameplate generating capability (P) becomes:

$$\frac{S_{incr}}{P} \approx Z_o r\alpha + \frac{Z_o \alpha^2 P}{2S_o}, \qquad f = \frac{s}{S_o} << 1 \qquad (9.4)^8$$

where Z_o is the original system z-statistic that meets reliability criteria. Take the small project example of project B previously described. The schedule error of 3 MW and nameplate capacity of 10 MW sets $\alpha = 0.3$ with a correlation (r) of 0.2 in a system where the z-statistic is 2.0. Then the first term of equation (9.4) is $2.0 \times 0.2 \times 30\% = 12\%$ of nameplate capacity.

The second term is less significant than the first where $r > f$, and is not dependent on the correlation factor. In this case it adds an additional 0.9% of capacity for a total of 12.9%, which compares with 13% calculated exactly from equation (9.2). A different, less diverse project with a correlation of 0.4 would double the first term (24% from 12%), but leave the second term (0.8%) unchanged. This shows the importance of correlation factors, but the variability of the output (α) as defined by schedule errors is nearly equally important.

9.4.6 Dependence on order

The foregoing analysis is based on determining how an existing system reserve requirement changes with a single addition to it. The analysis can be used successively to determine the effects through time of many additions. It may be worth noting some of the less intuitive results.

For example, if project A is followed by an identically sized project with a similar correlation, the incremental reserve requirement is less for the second added project than the first. The reason is that while the first project standard deviation of 30 MW out of 75 MW in the existing system for an f-factor of 0.40, the second added project is 30 MW out of 75 MW + 22.3 MW = 97.3 MW for an f-factor of 0.308 and a fractional incremental reserve requirement of 0.104 (equation (9.2)), or 0.104×97.3 MW = 10.1 MW of incremental reserve required compared to 22.3 MW for the first project A. The reserve requirement as a fraction of nameplate capacity drops from 22.3% to 10.1%.

Of course, if the schedule errors on the second project are closely correlated with those of the first, the overall correlation would probably be higher than 0.2. For example, if the second project has an overall correlation of $r = 0.5$ instead of 0.2, the increased reserve requirement goes from 10.1 to 17.9 MW. In any event, it is likely that a second project, even

---

[8] The derivation of this formula assumes that the proportional increase in reserve requirement is equivalent to the proportional increase in schedule error standard deviation—only identically true for normal probability distributions.

one closely correlated to the initial project, is likely to have a smaller effect on reserve requirements.

9.4.7 Real data and the inconstancy of the z-statistic

The formulation in equation (9.2) is general and precise for determining the change in standard deviation when the schedule error terms of an existing system are joined with another set of schedule errors. It is not dependent on the distribution of errors—it is accurate whether or not the errors of either set follow a normal distribution. However, the fractional increase in standard deviation precisely implies an identical increase in the reserve level only when both sets of numbers follow normal distributions. In practice the error introduced by assuming normal distributions may nevertheless be relatively small.

For example, the data in Figure 9.5 represent a distribution of schedule errors from a combination of load and wind errors. The standard deviation of load schedule errors is 110.9 MW, the standard deviation of wind schedule errors is 69.3 MW, and the correlation between the data sets is $r = -0.031$. From this, the fractional change in standard deviation from the original load series can be accurately computed from equation (9.2) as 16.26%, for a combined standard deviation of 129 MW.

Using the original z-statistics associated with 95% confidence intervals and applying the fractional increase in standard deviation to estimate new reserve requirements returns −271 and 219 MW. These estimated values compare with the directly measured confidence interval of −253 and 214 MW from Figure 9.5. The normality assumption introduced errors of 7% and 3% respectively in this particular example. Given that neither of the original distributions was normal, and the added schedule errors were a significant fraction of the existing ones, this lends some confidence for the overall approach.

9.4.8 Conclusion

Potentially useful results of the analytic approach that may help inform rate-setting include the following:

1. Incremental projects tend toward smaller reserve requirements as a fraction of nameplate capacity.
2. Small projects generally cause smaller incremental reserve requirements as a fraction of nameplate capacity than larger ones.
3. Reserve requirements for small projects are proportional to the product of correlation and variability/uncertainty.
4. Using the analytical techniques provided a reasonable estimate of incremental reserve requirement in a real data example.

9.5 SUMMARY

Integration costs can be determined fully, or in part in a separate analysis that may be used for establishing wind-specific costs for comparing wind generation value outside system models or for establishing tariffs to recover integration costs. The methods are generally similar to the valuation analysis in the previous chapter, and many studies employ chronological economic dispatch models (CEDMs) to determine stand-alone wind integration costs. Wind integration costs can also be calculated using methods that are partly or wholly separate from CEDMs. Costs associated with increased reserve requirements generally form the bulk of the wind integration costs. The effects of incremental projects on incremental reserve requirements can be estimated and show the relative importance of wind forecast accuracy, wind generation variability, and project size. Incremental wind integration costs for a particular project are not fixed, but dependent on the order in which the project is added to the system with respect to other wind projects, and system load growth.

Chapter ten

Wind Power's Contribution to Meeting Peak Demand

If the wind will not serve, take to the oars.

Latin proverb

An aspect of wind generation value that receives significant attention is the extent to which wind generation may be depended on to meet power system peak demand. Sometimes referred to as the 'capacity value' of wind generation, the answer is complicated due to several factors. First and foremost is lack of uniform agreement among power system planners over exactly what is meant by capacity contribution, too-frequent misunderstandings over the way in which conventional generators combine to reliably meet peak demand, and the fact that the contribution from individual generators depends on the other generators on the system. A variety of approaches to establishing the capacity value of wind have been used. The approach presented here is based on establishing the additional peak demand that can be met as a result of an incremental addition of wind generation. The chapter also explores some characteristics of wind capacity value (e.g. its dependence on other generation) and examples of wind capacity contribution studies are provided.

Capacity value of wind power has been addressed in a number of publications. In Castro & Ferreira (2001), methods for capacity value are described, and classified as either chronological or probabilistic. A range of methods for the calculation of capacity value are assessed by Milligan & Porter (2005, 2008). A generalized version of methods is presented in D'Annunzio & Santoso (2008), with the key innovation being a multi-state representation of wind power.

Some analysts adopt the view that the inconstancy of the wind, and the relatively larger probability that no wind generation is available during

Valuing Wind Generation on Integrated Power Systems. DOI: 10.1016/B978-0-8155-2047-4.10010-9

peak load periods, directly implies that wind generation does not contribute to meeting peak loads (Pavlak, 2008). However, the error of this view should be apparent in as much as numerous traditional specific generators experience unexpected outages at critical times—this does not mean that they provide no contribution to meeting peak demands through time. It is important to view the contribution to meeting peak loads in probabilistic terms to appropriately account for the availability and variability of wind (and other resources) at critical times.

Although it is not correct to decide a priori that wind generation contributes nothing to meeting peak demands, it is generally acknowledged that in most cases the contribution from land-based wind generation is relatively modest. It may be intuitively clear that the amount of capacity contribution depends on both the expected (i.e. long-term average) wind generation available during critical time periods (peak seasons of the year, days of the week, and hours of the day), and also the variability about that expectation. In other words, both the expected generation levels and variability about the expected levels need to be considered when determining capacity contribution. While this is equally true of traditional generating technologies, these methods are not commonly used to determine capacity contribution to other technologies, and may not be familiar to all analysts.

10.1 CAPACITY VALUE AND EFFECTIVE LOAD-CARRYING CAPABILITY[1]

Much of the material already presented relates to ensuring the ability of the power system to meet specific reliability criteria relating to balancing demand and generation by establishing sufficient quantities of reserve generation. Generation system adequacy is concerned with ensuring sufficient generating capability to meet the demand at any point—especially focused on times of the year and times of the day where demand historically peaks (e.g. hot summer days or cold winter evenings). Adequacy is achieved through the combination of generators on the power system, all potentially contributing to meeting overall demand for power. Capacity value can be defined as the amount of additional load that can be served due to the addition of the generator, without diminishing the likelihood of meeting system demand.

Capacity value has not historically been used by power systems as the primary tool used to establish and ensure system adequacy. Many systems continue to use methods that evolved over time to ensure system adequacy based on planning reserve margin (PRM) (Milligan & Porter, 2005). Planning reserve margin is based on the nameplate capacity of generators. Because the nameplate capacity of wind generation is not a good indicator

[1] This chapter borrows heavily from Keane, Milligan, D'Annunzio, et al. (2009). Capacity value of wind paper: task force on the capacity value of wind power. *IEEE Power and Energy Society*, submitted November 2009.

of the capacity value of wind generation, PRM cannot be used without relatively extensive modification. This situation has left many systems in something of a quandary as to the extent to which wind generation can be relied upon to meet system peak demand—if at all.

A seeming likely candidate measure of capacity value might be the expected level of generation from resources during the peak demand periods. This is different from nameplate capacity due to unplanned outages. However, actual availability will be less than the expected availability roughly 50% of the time. Power systems targeting higher levels of reliability will therefore account less than the expected generation availability. Other methods are needed to determine a capacity value based on target reliability levels.

The concept of capacity value has gained importance as reliance on wind generation has increased. Capacity value can be defined as a measure that assesses the extent to which wind generation can be relied on to meet peak demand. Best practices in assessing capacity value are dependent on probability-based reliability metrics. Several metrics for assessing service reliability exist that are variously used in determining capacity values. These include the loss of load expectation (LOLE), expected unserved energy (EUE), and the loss of load probability (LOLP). It should be noted that although these metrics are commonly referred to as reliability metrics, their values represent the extent to which a system fails to meet load, i.e. larger numbers represent less reliable power systems. For example, LOLP is the probability that the load will exceed the available generation at any given time. The metrics might more properly be thought of as 'unreliability metrics'.

While LOLP represents the probability an outage may occur, it provides no information as to the severity or duration of any potential shortfall. Another measure, EUE, is the average energy loss over some specified time period and is a measure of the combined effects of frequency, intensity, and duration of shortfalls. LOLE is a measure of the expected number of hours or days during which the load will not be met over a defined time period (e.g. over a year, or over 10 years).

The capacity value of a generator can be established using any of these measures—or potentially is determined using all of them to determine the maximum additional load that can be met without increasing any of the measures. Capacity value computed in this way may be[2] taken to be the increase in the effective load-carrying capability (ELCC) when a resource is added to a power system. Here, we will take capacity value as the change in ELCC, as determined using any or all of the three reliability metrics (LOLE, EUE, LOLP) cited above.

---

[2] Nomenclature is not entirely consistent. Some usages of ELCC may refer to capacity value determined using one of the specific measures noted, or in some cases a different metric altogether.

Here again, care must be taken in applying these practices to wind generation without the equivalent treatment of traditional generating resources. The question of wind capacity value arises because it is abundantly clear that the nameplate capacity (maximum generating capability) of wind projects is not indicative of the extent to which wind contributes to meeting peak load and relegating the PRM technique inadequate. It is a fair and obvious question, but it is less whether that the same question and arguably the same treatment should be applied to traditional generation.

Just as nameplate capacity is not a complete indicator of wind capacity value, neither is it an accurate measure of traditional generators' contributions to meeting peak demand. For example, generators with large forced outage rates clearly contribute less than more reliable generators. Similarly, a planning reserve margin representing fewer large generators does not represent the same level of reliability as the same PRM made up of more numerous smaller generators. Ideally, the capacity value of all generators would be assessed on an equivalently fair basis as that proposed for wind generation. The methods recommended below are equally applicable to wind or other resources.

10.2 COMPUTING EFFECTIVE LOAD-CARRYING CAPABILITY

Determining ELCC is at heart a probabilistic exercise, necessitating some means of determining the probabilistic reliability metrics LOLP, LOLE, or EUE. It is necessary to first establish probabilities associated with meeting various levels of demand. The most direct method entails enumerating all combinations of generating capability levels. An example of such a process is shown in Table 10.1 for the case of five 50-MW generators where each generator has a 5% probability of being unavailable and a 95% probability of being available. The probability of each generating state is also shown. Tables enumerating all possible states of generator operation such as that shown in Table 10.1 are called capacity outage probability tables (COPTs). Every possible combination of operating and non-operating generating units is documented, with the associated probability of the system finding itself in such a state.

Table 10.1 shows that the most probable state is all generators available (last row of the table), with a probability of 77.4%. By implication, these generators would be unable to meet a load of 250 MW almost one-fourth of the time. On the other hand, the system can meet 200 MW of load (one unit inoperable, four units operable) a total of 93.6% of the time—still failing a modern reliability criterion that might be set at 95% or more. These generators could meet 150 MW of load (at least three generators available) 99.7% of the time.

Data such as that presented in Table 10.1 can be used to calculate the reliability of meeting loads with a given system. Figure 10.1 shows just

TABLE 10.1 Capacity Outage Probability Table (COPT) for Five 50-MW Generators in Which Each Generator is Available 95% of the Time, and Unavailable 5% of the Time*

Generator 1	Generator 2	Generator 3	Generator 4	Generator 5	Generation Level	Probability Level
0	0	0	0	0	0	0.00003%
0	0	0	0	1	50	0.001%
0	0	0	1	0	50	0.001%
0	0	0	1	1	100	0.01%
0	0	1	0	0	50	0.001%
0	0	1	0	1	100	0.01%
0	0	1	1	0	100	0.01%
0	0	1	1	1	150	0.21%
0	1	0	0	0	50	0.001%
0	1	0	0	1	100	0.01%
0	1	0	1	0	100	0.01%
0	1	0	1	1	150	0.21%
0	1	1	0	0	100	0.01%
0	1	1	0	1	150	0.21%
0	1	1	1	0	150	0.21%
0	1	1	1	1	200	4.07%
1	0	0	0	0	50	0.001%
1	0	0	0	1	100	0.01%
1	0	0	1	0	100	0.01%
1	0	0	1	1	150	0.21%
1	0	1	0	0	100	0.01%
1	0	1	0	1	150	0.21%
1	0	1	1	0	150	0.21%
1	0	1	1	1	200	4.07%
1	1	0	0	0	100	0.01%
1	1	0	0	1	150	0.21%
1	1	0	1	0	150	0.21%
1	1	0	1	1	200	4.07%
1	1	1	0	0	150	0.21%

(*continued on next page*)

TABLE 10.1—cont'd

Generator 1	Generator 2	Generator 3	Generator 4	Generator 5	Generation Level	Probability Level
1	1	1	0	1	200	4.07%
1	1	1	1	0	200	4.07%
1	1	1	1	1	250	77.38%

** Generator availability is indicated by a 0 or 1, with 1 indicating the generator is available and 0 indicating an outage for that generator. Each possible state of the system is enumerated, along with its probability of occurrence and corresponding total available generation.*

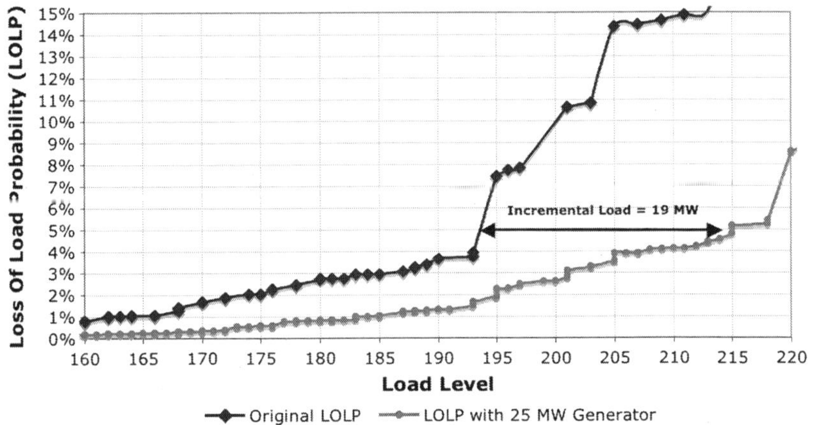

FIGURE 10.1 An example of LOLP as a function of the load level. The top line represents a base system and the lower line the same system with an additional 25 MW of generation. The figure shows that the added generator brings 19 MW of incremental ELCC at the 5% LOLP level.

such an analysis. The vertical axis shows the LOLP associated with meeting the load levels shown on the horizontal axis. The top curve represents the reliability associated with a system of eight generators, and the lower curve represents the same system with an additional 25-MW generator. From the figure, it is clear that the LOLP drops at any given level of load with the incremental generator.

In the example illustrated in Figure 10.1, the added 25-MW generator increased the ELCC by 19 MW at the 5% LOLP level. In other words, the added generator has a 'capacity value' of 19 MW as that term is taken to mean here. There are some less intuitive characteristics of this definition of capacity value that should be noted. For example, the capacity value is dependent on the accepted level of LOLP. While the 25-MW generator contributes 25 MW of capacity value at the 5% LOLP level, its contribution at other levels is different—dropping to about 12 MW at the 3% level in this example.

Another less intuitive result is that identical units may have different incremental ELCC contribution depending on the order added. For example, a system composed of a single traditional generating plant with a forced outage rate (FOR) of 7% would be deemed to have no capacity value in meeting a 5% LOLP standard. The addition of a second identical unit would lead to a system ELCC of 50 MW (at least one generator operating would occur 95.1% of the time). While the first unit would be deemed to have no capacity value, the second identical unit would have a full 50 MW capacity value.

Despite these difficulties, capacity value, defined as the incremental change in ELCC, is a useful concept and correctly recognizes the incremental load-serving value of an incremental generator. The same basic methodology is equally applicable to wind generators and more traditional generating technologies.

In large systems made up of many generating units, the COPT can become very large and difficult to compute. Table 10.1 represents a system where generators are either operating or not operating. In practice, generators often have a multiplicity of states in which they are partially operable at various levels and probabilities.

Applying a COPT state-enumeration approach with wind generation is perhaps especially challenging. The availability states of wind generation—essentially all the generating values that can be taken on by a wind project—are uncountable. They can be 'binned' (e.g. establish the probability that the wind generation is between 22.0 and 23.0 MW), but the table becomes difficult to develop and multiple tables may be necessary in order to capture the probability levels as they may change by season and hour of the day.

Employing the ELCC method is not dependent on the COPT approach. Other probabilistic methods exist. Stochastic methods that sample generating states can be used. These methods select random numbers from distributions to represent possible individual generator states. Total system availability is taken as the sum of the individual availabilities. Multiple samples are taken to build up a distribution of the total generator availability. In this way it is not necessary to develop a table of every possible combination of generating states—a task that can present practical problems while including states of no practical significance. The accuracy of stochastic approximation is a function of the number of samples taken. This approach is often simpler to implement and is potentially a more straightforward way to represent complex availability states that are characteristic of wind generation.

Figure 10.2 compares the exact COPT results to those obtained from a stochastic sampling approximation. The histogram of generating states for

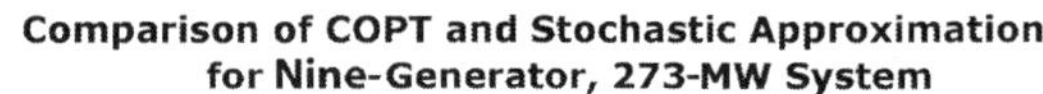

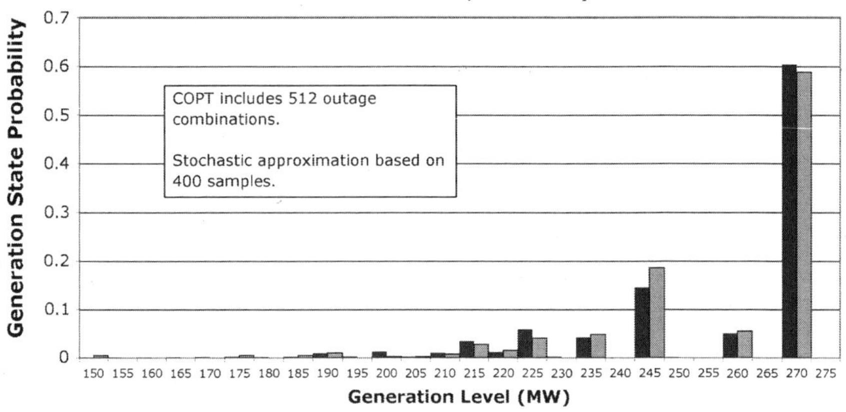

FIGURE 10.2 Comparison of the exactly determined histogram of a nine-generator, 273-MW power system using a COPT, with a 400-sample stochastic approximation. The accuracy of the stochastic approximation can be improved as necessary by selecting more samples.

the same system of generators is shown in the lower curve of Figure 10.1. That system includes nine generators having a total of 512 possible states. The stochastic approximation in the figure shows the results from just 400 samples. If a tenth generator were added to the system, the COPT table would double in size, whereas the complexity and size of the stochastic model would increase by just 11%. Changing a COPT to reflect additional generating states (i.e. not just on or off) would also necessitate major changes and increases in the size of the table, whereas the changes to the stochastic model would be very modest. For large, complex systems, employing stochastic models is often easier and more cost-efficient to implement.

In addition to the COPT and straight stochastic methods described above, chronological models may be used. Chronological models are necessary where the time evolution of processes is important. For example, the availability of hydro energy at any particular time is dependent on how the hydro generation was used at earlier times because generating capability depends on reservoir storage levels. Wind generation and loads also have important time-sequential properties that are not captured by the time-independent statistical analyses implied by the COPT and stochastic approaches. Chronological models can use stochastic methods where each sample represents a possible evolution through time in which generator availability and loads are sequentially represented. A limitation of the chronological approach is that each run of the model represents a single point on the curves shown in Figure 10.1. Multiple model runs in an iterative process may be necessary to determine the incremental ELCC.

Regardless of which method is used, the process for determining capacity value is the same:

- Determine the ELCC at some target reliability level
- Add a generator and compute the new ELCC at the same target reliability level
- Assign the capacity value as the change in ELCC.

10.3 WIND CAPACITY VALUE CHARACTERISTICS

Capacity value, defined as incremental ELCC, can be expressed in a generalized formula (Dragoon & Dvortsov, 2006) under certain limiting conditions for large systems adding relatively small amounts of generation. Despite limitations and assumptions, the resulting relationship is useful in understanding some of the less intuitive characteristics of capacity value that are likely to have more general applicability. The incremental ELCC for a generator that is small compared to the existing system (the capacity value) may be approximated by:

$$\text{Incremental ELCC} \approx R' - Z_o \sigma_{R'} \left(\frac{\sigma_{R'}}{2\sigma_s} + \rho_{R',S}\right) \qquad (10.1)$$

where R' is the expected on-peak generation from the added generator, $\sigma_{R'}$ is the standard deviation of the on-peak generation from the added generator, σ_s is the standard deviation of surplus generating capability for the existing system, Z_o is the ratio of σ_s to the expected surplus of the existing system at its target reliability level, and $\rho_{R',S}$ is the correlation during on-peak hours between the generation from the added resource and the existing system surplus generation.

The equation is developed from viewing the difference between demand and generating capability of the existing system (i.e. surplus generation) as a probability distribution. Equation (10.1) is the condition that maintains the initial system's level of reliability as represented by Z_o.

The first term in equation (10.1) is the expected on-peak generation from the incremental generator. As discussed previously, the expected generation level should not generally be viewed as the capacity value of the incremental generator because there is roughly a 50% chance that a lower level will be available. The second term in equation (10.1) is the reduction in capacity value due to the variability ($\sigma_{R'}$) of the incremental generator around the expected level. Notice also that the deduction is larger for higher values of Z_o, indicating that as we demand greater and greater reliability levels, the capacity value we accept from the incremental generator diminishes. This latter effect is consistent with the example system in Figure 10.1 that shows generally diminishing levels of incremental ELCC as the system reliability increases (lower levels of LOLP).

The terms inside the brackets in equation (10.1) show that the reduction in capacity value from expected generation is also dependent on the variability of added generator with respect to the existing system, and the correlation of the generation with the existing system variability. Traditional generators generally may be expected to have zero correlation with existing system surpluses, and the second bracketed term would be zero.[3] This is not generally true of wind generators where an incremental wind generator may be correlated with another relatively nearby generator.

Wind generation added to a system with existing wind where the generation levels are correlated with one another has diminishing capacity value. In other words, the relative capacity value of wind projects built out in an area of existing wind projects will likely diminish. Alternatively, wind projects built in geographically and climatologically diverse areas will generally return a larger portion of their expected generation as capacity value.

10.4 CASE STUDIES

Included below are summaries of wind studies performed by various entities seeking to determine capacity values for wind. These studies were performed by agencies in the US states of New York and Minnesota, Germany, and Ireland. The New York and Minnesota studies based capacity value on incremental reserve requirement as described above. The German wind study adopted an availability criterion as its definition. The Irish study sought to equate the capacity value of incremental wind generation to that brought by conventional generating technologies.

10.4.1 State of New York

The New York State Energy Research and Development Agency (NYSERDA) sponsored a study of the effects of wind energy on the New York power grid (Clark et al., 2005) prepared by GE Energy Consulting. The study's objective was to assess the effective load-carrying capability of future wind resources in the state of New York. It employed General Electric's MARS program, a chronological stochastic model that generally follows the incremental ELCC approach to establishing wind capacity value. MARS incorporates a representation of transmission system limits that simulates the ability to physically deliver power to the load centers.

Historical New York Independent System Operator (NYISO) hourly load data for 2001, 2002, and 2003 at different buses were used. The peak load for the period of investigation was 30,982 MW. The study

[3] Generator availability from physically separate generators is commonly assumed to be uncorrelated. This may not always be the case. For example, outages caused by temperature extremes may be correlated if distant generators are subjected to large-scale weather events.

incorporated 3300 MW installed wind capacity, of which 600 MW were taken to be located offshore. Historical meteorological data for the same years were used to create hourly wind power generation time series at different sites. Time-synchronized data for loads and wind generation were used to ensure proper correlation of the time series.

ELCC was calculated based on LOLE in consideration of physical transfer limits on transmission between pairs of interconnected areas. Historical loads for 2001 and 2002 were adjusted to reflect peak demand in 2008. The analysis showed most of the capacity value coming from the 600 MW offshore generation. An important transmission constraint limited the capacity value of wind sites in the western part of New York.

For 2001, the increased ELCC for the 3300 MW of wind generation was estimated to be 270 MW, i.e. 8% of nameplate capacity. Removing the transmission constraints improved capacity contribution to 720 MW (22% of nameplate capacity). Overall, the onshore ELCC is about 10% of nameplate capability compared with roughly 30% annual capacity factors. This result was primarily due to lower capacity factors during on-peak periods. Offshore wind capacity contribution neared the overall capacity factors of about 40% of nameplate capability.

10.4.2 State of Minnesota

The Minnesota Public Utility Commission ordered a study of power system impacts associated with meeting 20% of electric demand with wind generation. The study was performed in 2006 (Zavadil et al., 2006) and examined levels of wind generation corresponding to 15%, 20%, and 25% of forecasted 2020 demand: 3441, 4582, and 5688 MW respectively. Existing generation was augmented to meet a reliability criterion of LOLE of 1 day per 10 years in 2020.

Wind generation was represented as negative load and the analysis was conducted for three samples of the year 2020, where the hourly wind and load patterns were based on the historical years 2003, 2004, and 2005. Historical wind speed generation for assumed project sites was based on the MM5 numerical weather prediction model using historical weather data and 4-km grid sizes. Model output was taken for 152 grid points at 5-minute time resolution intervals. Generation levels were developed from wind speed using empirically derived relationships.

The LOLE analysis was performed using General Electric's MARS chronological stochastic dispatch model to assess system reliability based on LOLE. Incremental ELCC was computed for each of the three penetration levels. The effective capacity of wind generation varied significantly depending on the historical sample year. Wind ELCC for 15%, 20%, and 25% wind penetration ranged from approximately 5% to just over 20% of nameplate capacity.

10.4.3 German study

The German government targets 12.5% of electric energy to be generated from renewable resources through 2010, increasing to 20% by 2020. Pursuant to those goals, the Deutsche Energie-Agentur (DENA) undertook a wind integration study that was carried out from 2003 to 2005 (DENA, 2005). The study included an estimate of wind's contribution to the 'guaranteed capacity' at the time of winter peak. Guaranteed capacity was taken to be the amount of capacity that power plants can deliver at a defined probability level. This is a different definition of capacity value than the adopted definition here of incremental ELCC. A significant difference is that it is not based on load data. Nevertheless, it is another approach to capacity value that is considered by some analysts and is included as a widely quoted capacity value analysis.

Wind power availability was based on measurements of up to 220 wind farms covering up to 10 years of data. Different probability functions resulting from different time frames (coldest days, winter peaks, winter months) and different spatial distribution scenarios were also analyzed. The assumed acceptable availability level was set to 99%, meaning that the accepted capacity value was that available with a certainty of 99% of the time during the peak months examined.

It should be noted that most conventional generating units exceed 1% outage rates and would register as zero capacity value under this definition. The study estimated that 5–6% of the 36,000 MW of nameplate generating capability expected to be online in 2015 would meet the 99% availability standard.

10.4.4 Irish study

Eirgrid undertook a study of wind effects on the Irish national power grid (ESB National Grid, 2004). The study examined the impact of wind power generation on the conventional power plants. Ireland's peak system load in 2004 was approximately 4500 MW. Capacity value of wind generation was taken to be equal to the amount of conventional capacity that can be omitted while maintaining system LOLE standards. This definition of capacity value varies slightly from the definition of capacity value as the incremental ELCC. Alternatively, the Eirgrid study defined capacity value in terms of equivalent conventional generating units. Such a definition may have practical value, but as pointed out above the capacity value of conventional generating units varies by outage rates and unit size, making the comparison necessarily relative.

Hourly wind power generation was modeled as negative demand and added to the hourly load profile based on similar historical time horizons. Measurements from 18 onshore wind projects and one offshore project for 2001 served to represent the wind power profile. Capacity value was taken based on an LOLE reliability target of 8 h in one year.

The study showed incremental capacity value falling (approaching zero) as the amount of wind generation on the system increased. Capacity values were calculated based on power system peak demand of 5000 and 6500 MW. At 500 MW of wind capacity, the 5000-MW system resulted in a wind capacity value of 34% of installed capacity. With 1500 MW of wind, capacity value decreased to 23% of installed capacity. For the 6500-MW system wind capacity values were found to be 22%. Adding 3500 MW to the 6500-MW system resulted in a capacity value of 14% installed wind nameplate capacity.

Both the Minnesota and New York studies cited above developed incremental ELCC values specific to historical wind conditions. For planners seeking to include some specific capacity value in their studies this may not be satisfactory. Just as conventional resources are not evaluated based on performance in specific historical years, associating capacity value of wind projects with specific historical years provides little insight. Equation (10.1) suggests that year-to-year variability is a contributor to overall capacity value and should be included. Practical difficulties with sampling more than a few historical years has probably led to focusing on year-specific ELCC values. Improvements in analytical techniques may improve upon the current techniques and move toward ELCC values independent of historical periods.

10.5 SUMMARY

The extent to which wind generation contributes to meeting peak demand is broadly termed capacity value. Although no generally accepted definition of capacity value has been adopted, best practices appear to be to define capacity value as incremental effective load-carrying capability (ELCC). ELCC is defined as the peak demand level that can be met by a power system at a given level of reliability. Reliability is measured in terms of probabilities of meeting all demand, and common probabilistic metrics used for computing ELCC include loss of load probability (LOLP), loss of load expectation (LOLE), and expected unserved energy (EUE). Incremental ELCC is determined by starting with a base system meeting the target reliability metric and adding the incremental resource and sufficient load to bring the system back to its target reliability level.

A generalized formula for a somewhat idealized case suggests incremental ELCC depends on the expected on-peak generation levels reduced by terms representing both the variability of the on-peak generation and the correlation with existing system generation net of load. Several studies have been conducted to quantify capacity value using a range of methods. Two studies, done for the New York Independent System Operator (NYISO) and for the Minnesota Public Utility Commission, analyzed wind capacity values based on the incremental ELCC method recommended here. Those studies found wind capacity values ranging from 5% to 20%, with one study showing that offshore wind project capacity values approached their 40% capacity factors.

REFERENCES

Castro, R., & Ferreira, L. (2001). A comparison between chronological and probabilistic methods to estimate wind power capacity credit. *IEEE Trans. Power Syst., 16*(4), 904–909.

Clark, K., Jordan, G., Miller, N., & Piwko, R. (2005). *The effects of integrating wind power on transmission system planning, reliability and operations. NYSERDA, Tech. Rep.,* March 2005. <http://www.nyserda.org/publications/wind_integration_report.pdf>

D'Annunzio, C., & Santoso, S. (2008). Noniterative method to approximate the effective load carrying capability of a wind plant. *IEEE Trans. Energy Conv., 23*(2), 544–550.

DENA. (2005). *Planning of the grid integration of wind energy in Germany onshore and offshore up to the year 2020. Deutsche Energie-Agentur Dena.* March 2005. <http://www.dena.de/de/themen/thema-reg/projekte/projekt/netzstudie-i/>

Dragoon, K., & Dvortsov, V. (2006). *Z*-method for power system resource adequacy applications. *IEEE Trans. Power Syst., 21*(2), 982–988.

ESB National Grid. (2004). *Impact of wind power generation in Ireland on the operation of conventional plant and the economic implications.* February. <http://www.eirgrid.com/media/2004%20wind%20impact%20report%20(for%20updated%202007%20report,%20see%20above).pdf>

Milligan, M., & Porter, K. (2005). The capacity value of wind in the United States: Methods and implementation. *Electricity Journal,* no, 2, 9199–9204.

Milligan, M., & Porter, K. (2008). *Determining the capacity value of wind: An updated survey of methods and implementation.* Houston, TX: Windpower.

Pavlak, A. (2008). The economic value of wind energy. *Electricity Journal, 21*(8).

Zavadil, R., et al. (2006). *Minnesota Public Utilities Commission statewide wind integration study. NYSERDA Tech. Rep.* <http://www.puc.state.mn.us/portal/groups/public/documents/pdf_files/000436.pdf>

Chapter eleven

Effects of Markets on Wind Integration Costs

If wisdom were on sale in the open market, the stupid would not even ask the price.

Author unknown

Energy is routinely traded between suppliers and consumers in order to maximize the economic efficiency of power system operations and to balance supply and demand to minimize reliance on expensive generating reserves. Energy trading takes place under a broad range of market rules and regulations that have variously been established around the world. At one end of the spectrum, power companies may trade under voluntary bilateral agreements containing all the specific characteristics of the transaction (e.g. quantity, time, and place of deliveries) as mutually agreed between the trading partners. At the other extreme are structured mandatory markets with rules specifying terms and conditions of regular market auctions conducted by a centralized entity. Market rules can have large effects on the overall value of wind energy. Important market parameters affecting wind value include (DeMeo et al., 2009):

- **Market size/access.** The primary value of markets is to allow a balancing area or utility-owned power system access to diverse loads and generators in neighboring systems. If transmission access is limited, or the potential trading partners are small or non-existent, reserve requirements are higher.
- **Scheduling rules and imbalance settlement.** Operating period scheduling rules that allow shorter lead times for submitting schedules and more frequent scheduling (e.g. 10 minutes versus hourly) reduce overall schedule errors that contribute to the need for holding expensive generating reserves. Rules relating to financial settlements of operating period imbalances can also significantly affect the value of wind energy.
- **Ancillary service requirements and charges.** Much of this work is dedicated to establishing the costs of providing ancillary services for

Valuing Wind Generation on Integrated Power Systems. DOI: 10.1016/B978-0-8155-2047-4.10011-0

wind energy. However, the actual charges levied in any given market may or may not reflect the cost of integrating wind. At this time, many balancing areas do not have wind-specific charges. In areas where wind-specific charges are in place, they may or may not have been accurately determined given the complexities of accurately computing them as set out in this book.

- **Participation in redispatch.** A key factor in how efficiently a power system is able to respond to wind and load changes is the extent to which generation is allowed to participate in responding to relatively rapid changes in net system balancing requirements. Transmission system operators vary in the extent to which the intra-hour balancing requirements are met by the economic dispatch of available resources. Vertically integrated utilities tend to rely on their own resources. Conversely, some grid operators adjust generation sub-hourly on the basis of market auctions that both reduce the need for balancing reserves and the cost of providing balancing services.
- **Wind forecasting services.** Some markets require mandatory participation in wind forecasting services to develop wind schedules. Markets may charge for the centralized services, or conversely require wind generators to provide or pay for individualized wind forecasts. Rates designed to provide an incentive to the highest forecast accuracy are not in general use, and consequently mandatory forecasts may not represent best available technology.
- **Capacity valuation.** Organized markets may explicitly recognize the value wind energy brings in meeting peak loads—often termed 'capacity value'. The extent to which wind can be relied upon for meeting peak loads depends on a number of factors, including the effective size of the power system. Large well-integrated markets foster the effective use of wind generation for meeting peak load, and the market itself may offer capacity payments relative to the recognized capacity value.
- **Market incentives.** Some markets guarantee prices to renewable resources through 'feed-in tariffs', or may give priority to renewable generation during transmission congestion events. Conversely, in many systems the availability of transmission is limited and wind, as the last added resource, usually receives the lowest priority ('nonfirm') access to transmission. The existence of market incentives can greatly affect the valuation of wind generation from the producer's point of view.
- **Efficient use of transmission capability and construction cost recovery.** Wind resources tend to exist in less populated regions, often necessitating construction of new high-voltage transmission facilities. The costs of new transmission facilities are often too high for a single wind project to bear. Market rules influence how the costs of new

transmission facilities are recovered, and the efficiency of allocating existing transmission capability.

The effects of each of the above-mentioned market characteristics are examined in greater detail below, along with summaries of conditions in a sampling of existing markets. A compilation of how North American markets address each of these areas is available from the Utility Wind Integration Group (UWIG, 2009).

11.1 MARKET SIZE AND ACCESS

The overall value of wind energy is deeply affected by the power system into which it interconnects. At one extreme are isolated systems with barely sufficient flexible resources to meet the variability and uncertainty of their own demand. An example of that might be an isolated village or island without transmission access to neighboring power systems. Adding significant amounts of wind energy to such a system would likely entail significant changes to the existing power-generating facilities, or potentially the replacement of some baseload generation with more flexible generating capability.

In contrast, a system composed of a large number of generators capable of increasing or decreasing generation on short notice will largely accommodate moderate amounts of wind generation without any physical modifications or significant economic dislocations. The primary effect in such a system is the reduction in marginal fuel costs at times when the wind is producing energy. In large systems, integration costs as determined by methods described in Chapter 10 may not be significant.

In general, power systems can reduce the overall cost of accommodating wind generation by increasing the trading capabilities with its neighbors through large transmission interconnections or liberalized trading regulations. Movement toward electric market deregulation and the formation of regional markets has acted to generally lower the cost of integrating wind into power systems where such markets have formed (Kirby & Milligan, 2008).

Increased market size and access allows individual power systems to share both the available controllable resources, and the overall diversity that naturally exists among disparate load and wind generating centers. Diversifying wind and load through expanded markets brings the uncertainty and variability down overall as a fraction of the total load and wind. Formation of organized markets across load-serving entities has greatly facilitated the generator and load sharing that leads to lower integration costs. A Minnesota study (Minnesota Public Utilities Commission, 2006) showed that the effect of implementing a larger market roughly halved wind integration costs at comparable wind penetration levels as a fraction of load served.

11.2 SCHEDULING RULES AND IMBALANCE SETTLEMENT

Power generation and delivery schedules allow power system operators to prepare the power system in advance of the operating period in order to produce the most economically efficient use of available generation and to ensure sufficient resources to meet the expected demand. Markets vary in their approach to schedules. In isolated systems, or those without organized markets, scheduling requirements may be informal. In more formal markets, schedules are often submitted and updated at various times leading up to the operating period. It is relatively common for markets to provide for schedules in the day-ahead and hour-ahead timeframes. The preparation of day-ahead schedules facilitates decisions regarding starting up or shutting down power plants that may not be capable of responding to market conditions on an hourly basis.

Of most immediate concern to wind generators is the timing of the hour-ahead scheduling. Wind forecasts are most accurate for the immediate future, and rapidly decline in accuracy as time moves on. For example, wind forecasts based on persistence 2 hours in advance of an operating period may be half as accurate as schedules prepared 30 minutes prior to the operating hour. As is clear from the analysis of reserve requirements in Chapter 6, wind integration costs are closely tied to the accuracy of operating period schedule. Market rules requiring wind schedules to be submitted long in advance of the operating period can increase overall wind integration costs.

For example, some transmission system operators require schedules to be submitted 75 minutes prior to the operating period, while others require as little as 20 minutes lead time to provide generation schedules. There exists a tension between the desirability of reducing the lead time for hour-ahead schedules and the need for sufficient time to arrange overall balanced schedules for the operating period. The arrangement may include the need to enter into bilateral trades, or to conduct an auction to determine the least-cost dispatch of available resources. To the extent that market transactions can be concluded more quickly and efficiently, the net accuracy of wind schedules may be improved and reduce the cost of holding more expensive reserve generation units.

Differences between scheduled generation levels over an operating period and actual generation are common for most generating technologies, but the relative inaccuracy of wind forecasts means these differences are more common and more extreme for wind generators. Market rules are often put in place to discourage schedule imbalances—the extent to which the average generation over the operating period differs from the scheduled generation. Relatively extreme penalties are sometimes enacted for large scheduling errors, or errors that could constitute market gaming. Market gaming is the intentional mis-scheduling of generation levels for economic benefit. Relatively high penalties are levied on large schedule imbalances to prevent such activity. If imbalance penalty rules are not adjusted to accommodate the relative uncertainty of wind generation, they can become burdensome.

Given the relationship between wind schedule accuracy and wind integration costs, it may make sense to retain imbalance charges (not penalties!) to send the appropriate price signal to wind schedulers. Investments in wind forecast improvements have the ability to reduce wind integration costs and maximize the efficiency of the power system dispatch. Without an appropriate price signal, the wind scheduler will not see the value of incremental forecast improvements. Some transmission system operators have opted to centralize forecasting services. Providing price signals through charges levied on schedule imbalances is less important where forecasting services have been centralized to the entity providing imbalance energy and ancillary services.

The length of the operating period is also important in maximizing schedule accuracy and minimizing reserve requirements. Organized markets in North America allow redispatch of available generation ranging from 5 to 15 minutes (UWIG, 2009). Transmission providers in the USA outside the organized markets have traditionally used hour-long operating periods limited to no dispatch adjustments undertaken within the hour.

11.3 ANCILLARY SERVICE REQUIREMENTS AND CHARGES

Balancing requirements above those supplied through schedules and within-hour generation dispatch are met through holding reserve generation that may be recompensed through ancillary service charges. Not all markets levy wind-specific ancillary service charges. Most wind integration cost studies endeavor to determine the quantity and types of ancillary services required by wind generation. Existing tariffs may not fully capture the costs of providing those services. For reliability purposes, it is imperative that power system operators determine the appropriate types and quantities of ancillary services needed to accommodate wind on their systems. Reflecting those costs in ancillary service tariffs tailored to wind generation appears to be a growing trend. To the extent that those charges provide appropriate economic signals, they will be more likely to incentivize the most efficient build-out and operation of wind resources.

Ancillary services include generation reserves to support frequency stability, energy supply and demand balance over seconds and minutes, and supply restoration after outage events (contingency reserves). Although imbalance energy is covered as a separate category, there is a growing realization that there is an ancillary service component to supplying the energy imbalances over the operating period.

Just as some systems have been slow to understand and charge for the relevant ancillary services, others have jumped to lumping the costs together into roughly calculated 'wind integration costs'. Care needs to be taken to reflect actual costs in such tariffs, and not overstate them to the point that wind generation becomes broadly, and inappropriately, discouraged.

In North America, wind generators are often assessed charges for contingency reserves—generation available on short notice in the event of sudden loss of major power system components such as generators or transmission facilities. Whether wind generation should be charged the same as other generators is somewhat questionable, since an outage of a single wind turbine, or string of wind turbines, is unlikely to have an appreciable effect on power system reliability or the total contingency reserve requirement. Requiring wind generation to pay for such reserves in addition to ancillary services to cover wind variability and uncertainty may be duplicative and unnecessary. Alternatively, sudden loss of wind generation due either to high wind speed cutout or simply sudden loss of wind speed may be identified as qualifying contingency events (if suitably extreme) for which contingency reserves could be called upon, depending on regional reserve pool regulations.

Aside from generating reserves, ancillary services may include costs associated with voltage control. Voltage control is accomplished with a variety of reactive devices such as capacitors and inductors, as well as various active compensating devices. Grid interconnection agreements may require certain reactive capabilities on wind-turbine generators themselves. There may be additional requirements for active management of voltage control, especially for large projects located far from load centers. Costs incurred may be specially levied on wind generators if warranted.

11.4 PARTICIPATION IN REDISPATCH

Wind energy may be scheduled for delivery on day-ahead or hour-ahead bases. Once the schedule is set, the balancing area operator needs to adjust in real time to maintain a balance between supply and demand. Ancillary services in the form of generating reserves are available to help with this function or to relieve transmission congestion as needed. Balancing area operators may actively manage generators within the operating period to effectively reduce the overall need for generating reserves. Active management comes in the form of redispatch orders from the balancing area operators to individual generators, and sometimes to loads as well. Grid operators that conduct open auctions for sub-hourly balancing services reduce the burden on regulating reserves and increase the economic efficiency of responding to the balancing need.

Although wind generation may (when the wind is blowing sufficiently) respond to signals to either increase or decrease generation, wind redispatch is almost entirely relegated to responding to orders to decrease generation levels. Since wind generation has zero or perhaps negative production costs,[1] purposely reducing wind generation levels should ordinarily be undertaken only after other opportunities to reduce generation or increase load have been exhausted. Displacement of wind generation, or 'wind curtailment', primarily occurs for two reasons (Fink et al., 2009): to relieve transmission congestion, and during times when high wind generation coincides with low demand.

Controlling wind generation may be an inevitable result of wind generation becoming a significant portion of a power system's energy supply. Wind generation, even wind generation dispersed over wide geographic areas, will sometimes produce rapid changes in output. At present, the value of the energy sacrificed by controlling wind generation is far less than the cost of designing a power system capable of responding to these extreme but relatively infrequent events.

A range of strategies for effecting wind energy displacement are in use or under consideration, including (Fink et al., 2009):

- **Contractual rights to displace wind energy.** Bilateral power purchase agreements may contain provisions allowing the purchaser to reduce generation at the purchaser's own discretion, or under contractually specified conditions. If reductions are unlimited, contracts usually recompense wind generators for the lost production and any associated tax consequences.
- **Market-based curtailments.** Some markets allow wind to bid prices associated with generation reductions. Wind generation reductions are voluntarily undertaken when market prices come down to the bid levels. Market-based curtailments have been implemented in PJM, NYISO, and Nordpool.
- **Mandatory operating limits.** Such limits may include maximum coincidental generation levels, or limits to the individual project, or coincident increase in generation levels over a set time period. The ERCOT temporarily implemented daily operating limits in a portion of their system to address transmission congestion issues. Wind generation levels can also be limited after certain levels of reserves have been deployed. The Bonneville Power Administration imposed such a limit on wind generators in 2009.

Both contractual and market-based strategies may directly recompense wind generators for the lost generation. Mandatory operating limits

[1] Wind generation has no fuel costs, and low or zero variable operating and maintenance costs. There may be costs associated with reducing production due to lost revenues from power that may be contractually obligated on a bilateral basis, loss of the value of renewable energy credits not produced, or lost production tax credits. As a result, a wind generator may need a significant payment to remain economically whole when voluntarily reducing output. This can lead to negative market prices—i.e. offers to pay an off-taker for accepting energy at certain times.

generally place the cost of curtailments on the wind generators directly. In systems that levy wind-specific ancillary service charges, wind generators should realize a greater benefit through reduced ancillary service charges than would be levied without curtailments.

11.5 WIND FORECASTING SERVICES

Power system operators are gaining a greater appreciation of the importance of wind forecasts to reduce the need for and cost of holding reserve generation. Forecasts are used to establish generation schedules that are used prospectively to arrange the most cost-efficient use of available resources over each operating period. Wind forecasts are generally more accurate where more data are available. Individual wind projects may not have access to generating levels at competing projects. Since transmission system operators have access to data from all interconnected wind projects unavailable to the individual projects, there can be significant improvements made in forecasting by centralizing the function to the transmission operator. Centralized forecast costs can be spread among all participating wind projects.

North American markets currently employing or actively pursuing centralized forecasting include (UWIG, 2009): PJM, NYISO (New York), MISO (Midwest), ERCOT (Texas), CAISO (California), IESO (Ontario), and AESO (Alberta).

11.6 CAPACITY VALUATION

As suggested in Chapter 10, methods vary with respect to determining the appropriate amount of peak demand met by wind to use for planning or operational purposes. In addition to the uncertainty as to the level of 'wind capacity credit' is how markets may value and compensate for the capacity value that generators bring.

In determining an amount of capacity credit, it is impractical to use the methods in Chapter 10 for each new wind project as it connects to the power system. Instead, an approximation of the capacity credit is normally used, usually as some fixed fraction of the nameplate capacity of the wind projects. Table 11.1 shows data for North American organized markets (UWIG, 2009).

Compensation for capacity credit varies among organized markets as well, although many organized markets do not directly compensate for capacity contributions. In those markets, the capacity contribution is taken as a measure of reliability for planning and operational purposes. In some systems, individual utility companies make decisions regarding the need for resource additions based on reliability analyses that may take into account a capacity contribution for wind generation along with other generating technologies.

TABLE 11.1 North American Organized Markets: Determination of Wind Capacity Credit

Market	Capacity Credit Basis
PJM	Prospective: 13% of nameplate capacity Historical: 3-year rolling average over heavy load hours
NYISO	Wind capacity factor over heavy load hours in summer and in winter
ISO-NE	Median wind output over heavy load hours in summer and in winter
MISO	15% nameplate capacity
SPP	Wind generation at 85% excess levels on 10% highest load hours
ERCOT	8.7% of nameplate based on load-carrying capability analysis
CAISO	Set on wind generation at 70% exceedence level over select heavy load hours
Alberta ISO	20% of nameplate capacity pending further analysis
IESO	Based on historical median wind generation levels during peak hours

Organized markets operating without specific capacity payments implicitly rely on wholesale market prices to provide appropriate incentives for investors to undertake generation resource additions. Such systems with large penetration levels of wind may find wholesale prices broadly depressed due to the low variable cost of wind generation, and more volatile due to the variability of wind (Meibom, 2007). This may contribute to the need in organized markets for a more direct mechanism to incentivize new generation to meet peak demands. A listing of the status of capacity payments among North American organized markets may be found in UWIG (2009).

11.7 MARKET INCENTIVES

Incentives for wind generation exist in many parts of North America and Europe, established to broadly promote renewable energy to meet a range of policy objectives. Policy objectives include both environmental (to reduce pollution) and economic (to spur jobs and investments in rural areas). In the USA, some 37 states have established some form of minimum renewable energy requirement, voluntary standard, or energy goal (Chuang & Schwaegeri, 2009). In Europe, the European Union Energy Policy calls for meeting 20% of energy needs with renewable energy by 2020, driven largely by the desire to reduce greenhouse gas emissions that contribute to climate change.

11.7.1 Federal incentives

Market incentives provide the primary mechanism for achieving policy objectives of increased renewable energy production. Wide ranges of economic incentives facilitate meeting the policy objectives. Incentives may exist at both national and regional levels. Incentives in the USA include the following:

TAX INCENTIVES

The biggest federal incentive in the USA has been the production tax credit (PTC). The PTC was originally established under the Energy Policy Act of 1992 and allows roughly (periodically adjusted for inflation) $20 for each megawatt-hour of wind energy produced over the first 10 years of commercial operation. A change in the authorizing legislation in early 2009 allowed wind and other renewable energy projects to alternatively take advantage of an investment tax credit—convertible to an outright government grant.

Wind and certain other renewable energy resources are currently allowed accelerated depreciation schedules for tax purposes. The Modified Accelerated Cost-Recovery System (MACRS) allows resource owners to depreciate selected renewable energy resources over a 5-year period. MACRS was originally instituted in 1986 and expanded in 2008 to allow a 50% 'bonus depreciation' for projects starting commercial operations in 2008 and 2009. Projects were allowed to depreciate 50% of their value in the first year, with the remaining value depreciating over standard schedules.

LOW-COST RENEWABLE ENERGY BONDS

Public entities not paying taxes are able to take advantage of federally offered clean renewable energy bonds (CREBs). Originally established in 2005, CREBs allow public entity investors in renewable energy projects to issue bonds for which the interest is paid to the lender in the form of federal tax credits, effectively allowing the borrower access to zero or near-zero interest rate money.

RENEWABLE ENERGY PURCHASE REQUIREMENTS

Under the Public Utilities Regulatory Policies Act of 1978 (PURPA), utility companies are required to purchase power from certain 'qualifying facilities' (QFs) at their 'avoided costs'. Wind energy projects no larger than 80 MW are subject to becoming QFs.

RENEWABLE ENERGY PRODUCTION INCENTIVES

The US Federal Policy Act of 1992 established the Renewable Energy Production Incentive (REPI) for qualifying renewable energy generation. REPI was designed to match the federal production tax credit (PTC) for public entities that could not take advantage of the PTC.

Of the national policies in the USA, the PTC is probably the one measure most responsible for wind development. Its conversion to either ITC or government grant in early 2009 was instrumental in minimizing the strong downward pressure on wind development in the aftermath of the economic crisis in late 2008.

European countries employ a mix of feed-in tariffs (FITs) and minimum renewable energy production requirements, also called 'renewable portfolio standards' (RPS). Renewable portfolio standards establish minimum renewable energy content levels for sales to retail power customers. Feed-in tariffs establish set prices for renewable generation interconnected to the power grid. The level of the tariff may change by technology type and also evolve through time to act as an incentive for earlier generation while requiring higher efficiencies through time.

Spain and Germany, world leaders in the development of wind energy, rely principally on FITs for renewable energy incentives. In principle, FITs are similar to the PURPA QF incentive in the USA, in which utilities are required to purchase the output of renewable generation. However, European FITs are designed to make it economically feasible to develop renewable energy projects, whereas the PURPA QF system mandates prices set by utilities based on the most economically efficient competing resources. Nearly 20 European nations employ FITs.

An important aspect of FITs is the level of mandated charges. If set too low, the FIT will not be sufficient to encourage renewable energy development. On the other hand, if the level is too high it risks overcompensating projects that may not be the most economically efficient among the alternatives.

The UK and a handful of other European countries rely on renewable portfolio standards. An advantage of RPS over FITs is that the most economically efficient projects are more likely to be exploited first, at an overall lower cost to the power system as a whole. An important aspect of the effectiveness of the RPS revolves around enforcement or penalty mechanisms. Mandates to meet minimum requirements in and of themselves may not be effective. Utilities in the UK are required to meet the RPS requirement or pay a 'buy-out' of the shortfall in the amount of approximately €45/MWh (indexed from 2002).

Both FITs and RPS may succeed or fail to provide adequate incentives for meeting renewable resource development goals. If the level of the FIT is set too low, it may have no effect at all on resource development. Similarly, an ambitious RPS may also fail if there is an ineffective enforcement mechanism or the standard itself is set too low. The PURPA QF system is perhaps an example of an FIT where the tariff is too low to spur significant development. In 2009, the US Congress was considering RPS legislation that would largely count quantities of existing renewable resources that it would have been unlikely to spur any new development. On the other hand, FITs are responsible for the development in the leading renewable energy countries in Europe, and RPSs have also been the source of significant resource development in the UK and elsewhere.

11.7.2 Non-federal incentives

Sub-national entities have also been successful in providing significant and important incentives for developing wind energy, especially in the USA where no effective FIT or RPS exists at the national level. The various US states have implemented a broad range of incentives to promote renewable energy development. Among them are measures similar to tax incentives at the federal level, such as production and investment tax credits, reduced property taxes, sales tax exemptions, and accelerated depreciation. More than a dozen US states make available low-interest loans and grants as "clean energy funds" for renewable energy projects. More than half the states have established renewable portfolio standards under which much of the renewable resource development has taken place. Feed-in tariffs are less common and have not been as large a factor as RPS.

Other initiatives include meteorological tower loan programs and voluntary renewable energy certificate programs, under which utilities allow end-users the ability to pay a premium for renewable energy. Independent REC purchase programs also exist that sell directly to the end-users, sometimes over the Internet. Some work has been done at the state level to facilitate transactions under the PURPA QF law. This primarily takes the form of standardized contracts for smaller projects. The law's stipulation of avoided cost payments has hampered wider efforts, especially during periods of low to moderate fossil-fuel prices, which tend to depress utilities' avoided costs.

11.8 TRANSMISSION CONSTRUCTION COST RECOVERY AND EFFICIENT USE OF CAPABILITY

Sufficient high-voltage transmission capability is necessary to move the rapidly increasing levels of wind generation from relatively remote areas to populated load centers. The American Wind Energy Association identified transmission capability as the largest challenge facing significant increases in wind development in the USA (AWEA, 2008):

> Transmission of wind power from windy rural areas, where it is generated, to population centers, where it is demanded, is the wind industry's biggest long-term growth barrier.

The problem of transmission is twofold: the need for new transmission facilities and market structures for cost recovery, and the need to adopt market changes to foster more efficient use of existing transmission facilities.

11.8.1 Efficient use

Available transfer capability (ATC) is the incremental power that can be moved over transmission paths as determined by transmission owners.

Shortages of ATC are seen as major impediments to the expansion of wind energy in the USA (AWEA, 2008). In North America, ATC has traditionally been determined assuming that all contractual rights are simultaneously exercised at the maximum reservation rate. For example, if a path is rated at 1000 MW, and the owning entity has reserved 500 MW for its own uses and sold 250 MW to each of two other entities, it would determine that the remaining ATC is zero. This leads to under-utilized transmission facilities if the parties do not all use the line at their allocated levels on most hours.

Allocating transmission resources on the basis of maximum capability is a legacy of power systems designed primarily to meet peak loads on the hottest or coldest days of the year. Wind generation is primarily an energy resource, and a system designed primarily for transferring energy might allocate transmission capability differently than has historically been the case. One proposal for more efficient use of transmission facilities is a new 'conditional firm' product. Conditional firm offers the purchaser a limited right to transfer energy over transmission paths that may otherwise be fully allocated, but under-utilized. Conditional firm arrangements may provide firm transmission capability in all but a few months of the year, or hours of the day in specific months. During times when firm service is not guaranteed, service may be interrupted only as necessary when lines are otherwise filled. Conditional firm may also fix a maximum number of curtailment hours in a year or month.

Another opportunity for improving transmission usage is to implement congestion redispatch schemes that pay loads or generators (including wind) to change operations to prevent or reduce congestion on transmission paths. Although the redispatch entails additional costs paid to the participating generators and loads, the cost incurred can be a small fraction of the costs of new transmission-line construction.

Markets where transmission providers provide conditional firm service and congestion redispatch have the potential to substantially reduce the costs for new wind entrants by reducing the need for new transmission construction.

11.8.2 Transmission construction cost recovery

Barriers to construction of new transmission lines are considerable both in North America (where distances from wind resources to load centers are typically hundreds or even thousands of kilometers) and in Europe (where high population densities make siting especially difficult and expensive). Sources of funds and cost recovery for the expenditure of those funds are important challenges in constructing new transmission facilities.

Responsibility for financing transmission construction in the aftermath of deregulation has become confused in the USA. The original plan for independent developers acting under market forces to plan and construct needed new transmission facilities did not materialize. In its wake, regional planning organizations and government entities are beginning to band together to identify promising wind resource areas where resources and transmission can be developed with a minimum impact to the environment. It is still unclear how the plans now in development will be financed.

Regions where entities are actively and productively engaged in coordinating planning and financing new transmission facilities will have lower overall wind development costs than regions where transmission limitations relegate wind sited to areas where the wind resource is less optimal.

11.9 SUMMARY

Market structures have a significant role in determining the value of wind energy or, conversely, the cost of integrating wind energy into the system. Market size and liquidity, especially of sub-hourly energy trading markets, directly affect the cost of integrating wind power. Market rules are also important, especially the timeline for submitting hour-ahead schedules. Requiring schedules to be submitted 2 hours prior to the operating period results in the need to hold significantly more reserves than rules allowing significantly less notice (e.g. 1 hour or less) due to wind forecasting errors.

Direct subsidies, tax incentives, minimum production standards, voluntary renewable energy markets, and feed-in tariffs also affect the valuation of wind power. Charges for ancillary services can be an important cost factor that varies across markets as well. Less important factors include charges and requirements for wind forecasting services, recognition of capacity value (usually not large for wind), and participation in dispatch markets.

Finally, transmission planning, financing, and cost recovery mechanisms can have a very large effect on the overall development of wind resources in an area, and the efficiency with which those resources are developed.

REFERENCES

American Wind Energy Association (AWEA) (2008). *Wind energy for new era: An agenda*, November.

Chuang, A., & Schwaegeri, C. (2009). *Ancillary services for renewable integration*. CIGRE/IEEE PES joint symposium, July.

DeMeo, E., Porter, K., & Smith, C. (2009). *Wind power and electricity markets*. Utility Wind Interest Group, December. <http://www.uwig.org/WindinMarketsTableSept09.pdf>

Fink, S., Mudd, C., Porter, K., & Morgenstern, B. (2009). *Wind energy curtailment case studies*. National Renewable Energy Laboratory. October.

Kirby, B., & Milligan, M. (2008). Facilitating wind development: The importance of electric industry structure. *Electricity Journal, 21*(3), 40–54.

Meibom, P. (2007). In my view. *Power and Energy Magazine, IEEE,* Nov.–Dec., *5*(6).
Minnesota Public Utilities Commission (2006). *Final report—2006 Minnesota Wind Integration Study,* Vol. 1, November.
Utility Wind Integration Group (UWIG) (2009). *Wind power and electricity markets,* August. Available at: <http://www.uwig.org/WindinMarketsTableSept09.pdf>

Chapter | twelve

Enhancing Wind Energy Value

Engineering is the science of economy, of conserving the energy, kinetic and potential, provided and stored up by nature for the use of man. It is the business of engineering to utilize this energy to the best advantage, so that there may be the least possible waste.

William A. Smith, 1908

Institutional and physical attributes of power systems affect the value of the wind energy serving electric power demand. Smaller systems with limited ability to trade with neighboring systems will be less able to realize the full value of wind generation compared to larger systems or those with significant trading opportunities with other systems. Hence, the value of wind energy is not fixed, but depends on the attributes of the interconnected system, as well as the relative amount of wind on the system. It follows that steps may be taken to enhance the value of wind energy experienced by a particular system. This chapter reviews some of the factors affecting wind energy value and potential actions that can increase the value of wind generation in meeting power system demand.

Wind energy value relative to more traditional generation is primarily affected by both the need to hold reserve generation to accommodate wind variability, and the cost of providing balancing services as they are called upon to counter the variability and scheduling errors. These needs can be reduced in a number of ways, including sharing balancing needs and services across larger regions, and improving wind forecast skill. Enhancing wind energy value implies taking steps to reduce the need to hold reserve generation (e.g. through better trading practices and better wind forecasting), and broadening access to diverse generators (market size and liquidity and transmission access) able to respond most efficiently to the balancing requirement at any moment.

Valuing Wind Generation on Integrated Power Systems. DOI: 10.1016/B978-0-8155-2047-4.10012-2

12.1 REDUCING RESERVE GENERATION REQUIREMENTS

Reserve generation requirements are affected most directly by the accuracy of wind schedules and the effective size of the region over which balancing services for wind and load are to be provided. Figure 12.1 shows the dramatic influence schedule accuracy has on the need for maintaining reserves. Also, there is an important role served by expanding the diversity of sources of variability. Figure 2.6 shows how the relative variability of wind generation decreases with geographic area. An analogous effect holds for increasing the geographic region representing balancing area loads. In perhaps the most straightforward example of increasing load diversity, including loads from across time zones has the effect of reducing the extent to which load variations occur simultaneously. Broadening the diversity of wind and load either by expanding the boundaries of the relevant balancing area, or by sharing balancing services and needs across neighboring balancing areas, reduces the relative need to hold reserves.

Although geography physically limits the ability of some power systems to access resources in neighboring areas, systems sometimes isolate

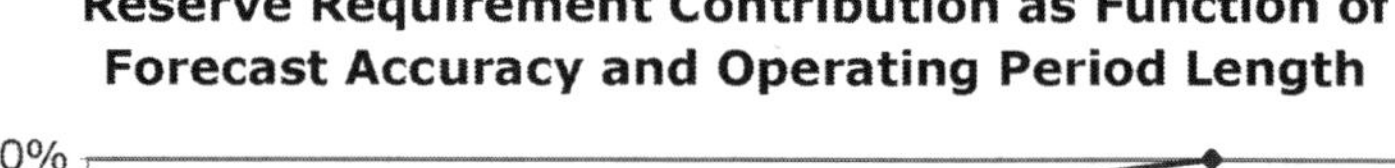

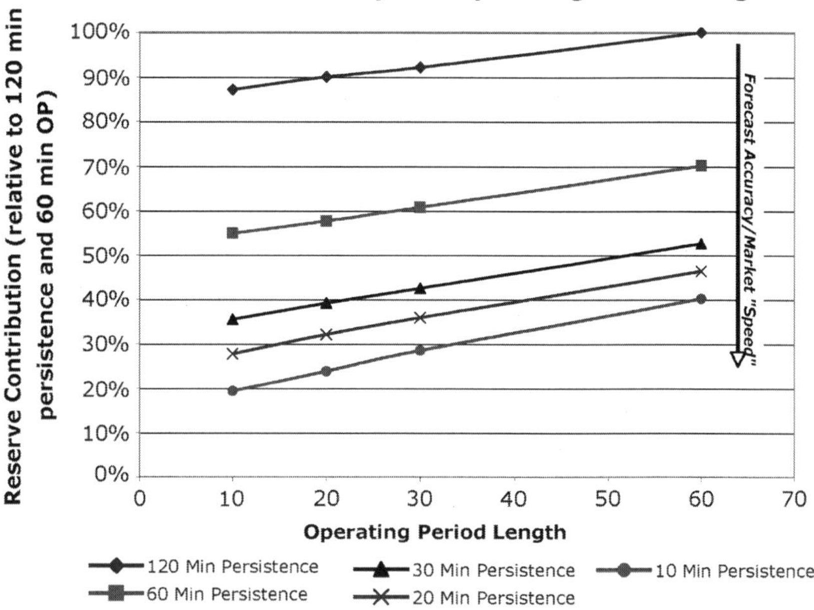

FIGURE 12.1 The relative effects of gate closure times ('market speed')/forecast accuracy and operating period length on reserve requirements. Balancing requirements are expressed in terms of the fraction of reserves relative to 120-minute persistence forecasts and 60-minute operating periods.

themselves institutionally through trading practices or transmission usage policies. Increasing access to flexible resources in neighboring systems can substantially reduce the need for carrying reserve generation. Reducing reserve requirements can be effected by:

- Improved wind forecasting
- Expanding sharing of balancing needs and services
- Shorter scheduling lead times and more frequent market transactions.

The relative importance of these measures is described below.

12.1.1 Improved wind forecasting

At the present level of wind forecast accuracy, the majority of the need to hold reserve generation is due to forecast error. Increasing wind forecast accuracy will directly increase the value of wind energy. Although day-ahead forecasts have been shown to be of very high value in some market structures, the most important area of improvement in balancing areas without intra-hour balancing markets is forecasts for the next 1–3 hours. It is likely that there is significant room for improving such forecasts through the application of more sophisticated modeling techniques in conjunction with more and better data. The present state of wind forecasting in the near-term, 0- to 90-minute timeframe is hardly better than using persistence forecasts. Near-term forecasting is a difficult problem, and commercially available forecasts have not been optimized for either power production or near-term forecasting purposes.

As methods for assessing the value of wind forecast accuracy improve, the value of forecasts is becoming clearer and providing economic incentives to improve near-term wind forecasting capability. In the USA, the National Oceanic and Atmospheric Administration (NOAA) is developing its rapid update cycle (RUC) and rapid refresh (RR) techniques that will provide more frequent and more current base data for numerical weather prediction models. Weather forecasting agencies are beginning to work more closely with the power industry to better tailor methods and data to the somewhat unique application to wind power forecasts. Of particular interest in wind valuation is the ability to predict regional wind ramping events (i.e. rapid and widespread increases or decreases in wind generation) and the accuracy of a particular forecast.

Figure 12.1 shows the relationship between forecast improvements and reserve requirements. Taken together with data illustrated in Figure 7.3, the data show a roughly one-for-one correspondence between forecast improvements and reductions in reserve requirements. In other words, a 10% reduction in NRMSE forecast error can, at least in the case of the data examined here, result in a 10% reduction in reserve requirement.

Even relatively modest increases in forecast accuracy can translate to large savings for power systems with high wind penetration rates. For example, a large system of several thousand megawatts of wind generation may require several hundred megawatts of reserve generation. A 10% improvement in wind forecast accuracy could therefore result in reducing the reserve requirement by several tens of megawatts. At a value of roughly $10/kW-month, the resulting savings might be worth several hundred thousand dollars per month. It is this kind of analysis that motivates efforts to improve near-term wind generation forecasting technology.

Although the benefits of improving forecasts are clear, the ability to produce them is less clear. Theory suggests that more and better data gathering and processing in near real time will lead to forecast improvements. However, there is no guarantee of success and a test case where a climatological region of complex topography is selected for detailed data gathering and modeling would provide needed guidance relating to the potential benefits of investments in forecasting.

12.1.2 Shorter scheduling lead times

Persistence wind forecasts are typically used for establishing reserve requirements, and are the most prevalent near-term forecasting methodology in use. As is illustrated in Figures 7.1–7.3, the accuracy of persistence forecasts drops rapidly as the forecast horizon ('look ahead' time) increases. One way to reduce forecast error and consequently the need for holding reserves is to reduce the forecast horizon—i.e. how long in advance schedules are required to be submitted to the grid operators and scheduling counterparties. In effect, the advance time, also referred to as the 'gate closure' time, directly affects schedule accuracy. Reducing gate closure times is equivalent to increasing wind forecast accuracy and reducing the resulting reserve requirement.

For example, if the gate closure time is 120 minutes prior to the start of the operating period, the persistence schedule would be the level of wind generation at that time. If gate closure occurs 60 minutes prior to the operating hour, the schedule would be set at the wind generation level at that time, resulting in a significant accuracy improvement.

The family of curves in Figure 12.1 shows how reserve requirements for persistence forecasts drop with reduced gate closure times. In the example of 120-minute versus 60-minute gate closures and 60-minute operating periods, the reserve requirement is just 70% of that required for 120-minute gate closures. Although the data illustrated in Figure 12.1 represent a specific data set, they strongly implicate the importance of both forecast accuracy and gate closure times in establishing reserve requirements.

The data illustrated in Figure 12.1 assume that reserves are held for all schedule errors over the duration of the operating period. Some markets allow generators to be adjusted within the operating period to provide some of the balancing services. In general, this is a less expensive means of dealing with wind and load variability on sub-hourly timescales than holding large quantities of reserves. Within-hour dispatch of generators is a significantly more efficient utilization of generating capability than maintaining reserve generation. Nevertheless, minimizing gate closure times increases forecast accuracy and reduces the amount and cost of intra-hour dispatch of generating units.

12.1.3 More frequent market transactions

Both gate closure times and operating period length affect the need to hold reserve requirements, as suggested by Figure 12.1. In markets such as PJM, MSIO, NYISO, ISONE, Nordpool, and the California ISO (and proposed for ERCOT), intra-operating period transactions occurring as often as 5 minutes substantially reduce the need to hold reserves relating to scheduling errors. Mechanisms to ensure sufficient bids and offers are available in the upcoming hour to provide balancing services may be necessary to ensure reliability while reducing the need to hold separate operating reserve requirements. Irrespective of the presence of enforcement mechanisms, reducing operating period length and implementing shorter-interval balancing markets outside fixed schedules serve to make more efficient use of wind generation.

The relationship between operating period length and reserve requirements reflected in Figure 12.1 may be realized to some extent for systems that currently use longer operating periods, or that allow for more frequent dispatch within longer operating periods. If operating periods are shorter, the reserve requirement is more directly affected.

12.2 EFFICIENT PROVISION OF BALANCING SERVICES

Balancing area services are most efficiently provided when balancing needs and service providers are shared over the broadest possible areas. Where transmission system operators exist that have combined previously separate balancing areas, provision of services will be most efficient—both because aggregating balancing service needs reduces the total requirement for balancing, and because entertaining a greater variety of resources enables the most economically efficient generators available to provide the needed service at any particular time. Finally, allowing auctions to provide balancing services helps ensure that the most efficient available generator is actually used to respond to the balancing need.

12.2.1 Wider sharing of balancing needs

Sharing balancing services and needs across broader geographic regions serves two functions. Balancing needs show relatively little correlation from balancing area to balancing area. Consider the uncorrelated fluctuations in demand (or demand net of wind generation) between two balancing areas A and B. The balancing requirement at any one moment in the two areas can assume one of four conditions:

1. Fluctuations in A and B are both positive (i.e. requiring incremental generation)
2. Fluctuations in A and B are both negative (i.e. requiring decremental generation)
3. Fluctuation in A is positive and in B is negative
4. Fluctuation in A is negative and in B is positive.

Under conditions 1 and 2, the fluctuations reinforce one another and the reserve requirement is additive. However, under conditions 3 and 4 the fluctuations tend to cancel one another. For uncorrelated systems, each of these conditions is equally likely, implying that half the time the aggregate balancing requirement is reduced simply due to aggregating the systems. The implication is that combining balancing areas reduces the total balancing reserve requirement.

Balancing needs over extended time periods may exhibit significant correlations, but the basic idea still holds—if there is sufficient transmission capability between balancing areas, aggregating balancing needs reduces the net reserve requirement. Figure 2.6 strongly suggests that wind generation exhibit very low correlations over periods on the order of several minutes and are the best candidates for sharing arrangements.

Reserve sharing pools have existed for many decades in all areas of North America where contingency reserve requirements are reduced by aggregating requirements among multiple balancing areas. Power pools in the east, which evolved into independent system operators (ISOs), have also reduced regulation requirements through aggregation, eventually merging multiple balancing areas together. In the western USA, vertically integrated utilities joined together to share balancing needs on the sub-minute regulation reserve timescale. Five balancing areas originally entered into an agreement in 2006 (NTTG, 2008) to net area control error (ACE) signals and allocate the net to participating balancing area operators. As suggested above, the netting reduces the need for reserve generators to respond to aggregated balancing needs. Similar arrangements among balancing areas over somewhat longer time periods would have a more pronounced effect on lowering balancing requirements relating to wind generation.

12.2.2 Incorporating a broader range of balancing generators

Allowing a wider range of generators and controllable loads to participate in providing balancing services is another way that the efficiency of providing balancing services can be improved. For example, consider two separate balancing areas, one dominated by hydro resources and another dominated by the presence of natural-gas-fueled generators. Hydro systems typically have excess incremental generating capability due to sizing the system to capture low-probability, high-streamflow-volume events. Environmental requirements may limit the ability for hydro generators to reduce their generation levels.

Conversely, thermal generation fueled by natural gas may have greater flexibility to reduce generation levels, but the maximum generating capability may be more significantly limited than in hydro systems. If these two systems are able to share incremental and decremental balancing services across balancing area borders, the economic efficiency can be significantly greater than if the two balancing areas are forced to independently provide both incremental and decremental balancing services.

Reaching a broader range of resources can occur by increasing liquidity of inter-balancing area trading and by consolidating balancing areas into larger balancing regions.

TRADING LIQUIDITY

Liquidity is the relative ease with which trading occurs among trading partners in a market. Trade volume is sometimes considered an indication of the liquidity of a market. Liquid markets are characterized by a large volume (number and size) of trades, and 'illiquid markets' tend to be characterized by fewer and smaller trades. Removing trade barriers between potential trading partners increases liquidity. Barriers can be either physical or institutional, or perhaps most prevalently a combination of both. Increasing liquidity allows more efficient use of available generation to provide balancing services.

Physical access to neighboring markets increases the pool of resources that may participate in providing balancing services and reduce the overall cost of accommodating wind generation. Access can be increased through institutional changes that facilitate trading, such as more frequent and shorter trading periods, or technological changes such as building additional transmission capability between markets or establishing automated trading platforms. Actions that facilitate the free transfer of energy increase the value of wind energy.

EFFICIENT USE OF EXISTING TRANSMISSION

Some markets allow for the exclusive purchase and reservation of rights over transmission facilities. For example, a transmission path between two points may have a maximum capability of 100 MW. If an entity purchases the 100 MW, no additional transactions may take place—irrespective of whether or not the purchasing entity actually uses the transmission. Other markets may allow for unused transmission capability to be sold separately, or may not have provision for exclusive transmission rights at all. Markets such as Nordpool do not allow fixed purchases of transmission capability. Trading partners may lock in prices through bilateral long-term financial instruments called 'contracts for differences', while physical trades take place in organized daily and hourly market bidding processes.

Balancing areas that allow for long-term purchases of transfer capability can result in under-utilized transmission facilities. Efforts to free up the under-utilized capability may require that unused transmission rights be resold through the balancing area operator. Without such requirements, entities may vie for competitive advantage by reserving transmission capability specifically to thwart more efficient usage by its competitors. Such systems are clearly suboptimal and do not lend themselves to the most efficient use of wind energy.

Certain technologies may also be employed to enhance the efficient use of transmission facilities. For example, devices such as phase-shifting transformers and solid-state equivalents can force prescribed power levels to flow over specific paths. This may sometimes be necessary to avoid overloading parallel paths and lead to an increase in the transfer capability of a particular transmission path. Better monitoring and control of transmission-line voltages, power flows, and phase angles to dynamically control the power system can also lead to more efficient utilization.

MARKET CLEARING PRICE AUCTIONS

Liberalized markets such as NordPool and the USA ISOs conduct supply and demand auctions in order to set prices for energy transactions. Energy service providers bid in their willingness to pay for energy typically in day-ahead and hour-ahead timeframes. Similarly, generators bid in prices reflecting their willingness to sell energy. Some markets allow some or all of the resulting balancing services to be determined in a similar way—through auctions. The most developed markets co-optimize the procurement of energy and ancillary services simultaneously, maximizing the generator's income and minimizing the total system cost. The effect of this is to reduce the cost of the balancing services and to limit the amount of reserves required.

AUTOMATING MARKET TRANSACTIONS

Automating transactions through either trade postings on automated trading platforms or automated auctions may vastly speed up the process of finding the most efficient allocation of generation to net demand. This can be of considerable importance to increasing the value of wind generation by effectively increasing the range of potential counterparties for transactions in near real time.

CONSOLIDATING BALANCING AREAS

Consolidation of vertically integrated utilities into large regional balancing areas served by a single transmission system operator (TSO), also called an independent system operator (ISO), reduces trading constraints and increases the economic efficiency of power systems as a whole. Access to a broader range of generators that may be able to increase or decrease generation in response to wind is an obvious advantage. In addition, balancing across a wider range of wind generators tends to decrease the net variability of the wind generators as a fraction of the installed nameplate capacity.

12.3 ACTIVE MANAGEMENT OF WIND AND DEMAND

Reserves are held to meet net variability and uncertainty of both demand and wind generation. Reserve levels are established based on relatively low-probability, extreme wind events. Typically such events involve the rapid and unexpected increase or decrease in wind generation levels. It is not possible for wind generators to substantially control their behavior during rapid loss of wind events, but several strategies may be employed to reduce the frequency and severity of some of the events that set reserve levels, effectively reducing the reserve requirement by the active management of wind generators.

Modern wind generators can limit their maximum generating capability at any moment. For example, if the wind generation rapidly rises over some time period to some predetermined level, consuming say 80% of the available reserves, a maximum rate of change could be imposed to ensure that any further increase occurs at a manageable rate. As another example, if forecasters foresee unstable and rapidly changeable wind conditions, they could limit the maximum capability of the wind generation to reduce the effect of both rapid rises and declines in wind generation. In systems where wind penetration levels are large, or where generation to provide balancing services is in short supply, it may be most economical to limit the wind generation at times rather than provide the balancing service from other generation.

An increasingly important way to provide balancing services is by accessing the storage capabilities inherent in many electrically powered devices. Such devices include refrigerators, freezers, building space conditioning (heating and cooling), water heaters (those with storage tanks), numerous industrial loads, irrigation pumping loads, and municipal water supply pumps. These devices share the characteristic that the consumption of energy maintains some flexibility with respect to the timing of the consumption. For example, the temperature of a building is regulated within certain acceptable limits. As long as the temperature is maintained within those limits, the space conditioning devices can be energized or de-energized as desired to provide balancing services. At present, few building space-conditioning systems possess the control and communication equipment necessary to use their storage capability, but they could be accessed by current technology.

Electric resistance heating elements are especially desirable as balancing devices as they respond virtually instantaneously, possess no moving parts to wear out, and are relatively inexpensive. Although electric resistance space heating is relatively uncommon, electric water heating with storage tanks is relatively common in some parts of North America. These devices could also be controlled to provide balancing services. The storage capability of a typical residential water heater is nearly equivalent to the storage capability of an electric vehicle battery.

Because these devices serve double duty providing space conditioning, hot water, or pumping services, the incremental cost to access the storage capability to provide balancing services is relatively inexpensive. The energy storage of a typical residential domestic hot water heater is equivalent to thousands of dollars of lithium-ion battery technology.

12.4 DEDICATED STORAGE TECHNOLOGIES

It may be natural to think of dedicated electric storage facilities and sometimes exotic technologies for storing energy as a means of maximizing the value of wind generation. Important storage technologies exist and have their place, but the cost of most of the technologies is daunting to say the least. In fact, most systems have relatively large storage capability in the form of reservoirs behind hydroelectric dams, natural gas fields and pipelines, and coal piles.

Indeed, it is these sources of storage that are already routinely employed to shape generation to load. Reservoirs are depleted during the day or seasonally to provide higher energy demands and refill at night or during spring freshets when demand is lower or streamflows are higher. Similarly, natural gas is withdrawn from production and storage fields during the day or through the winter heating months at greater rates than at night or off-season. In Europe it is relatively common to reduce coal generation at night, effectively using coal piles to store energy for when it has greater value during the day.

Just to give a sense of proportion, it might be helpful to compare the energy storage capability of a hydroelectric reservoir with the storage that might become available using electric vehicles. Franklin Roosevelt Lake behind Grand Coulee Dam in the USA by itself stores the energy equivalent of roughly 150 million electric vehicles with 30 kWh batteries in them.

Dedicated storage technologies include:

- **Pump storage.** Pump storage is achieved by using electric pumps to raise water from a lower level body of water to a higher level impound.
- **Compressed air energy storage (CAES).** CAES consists of electrically powered air compressors that pump air into large enclosures (usually underground chambers) that is later released through turbine blades to produce electricity—usually in conjunction with some additional combustion of fossil fuel to boost efficiency.

- **Battery storage.** Batteries are a very active area of investigation, principally for transportation applications. Electric motors are extremely efficient compared with thermal engines such as internal combustion technology.
- **Flywheel energy storage.** Flywheels are rotating wheels or cylinders that are spun up to high rotation rates using electric motors. The resulting kinetic energy can be stored at relatively high efficiencies on frictionless bearings in vacuum chambers. The energy is released by tapping the rotational energy through induction generators. Commercial units are now available on the scale of a few megawatt-hours of energy storage, sized primarily to provide fast-acting regulating reserves.

Each of these technologies represents assurance that energy can be stored at scale. At present, alternatives such as controlling wind generator output, modulating consumer demand, and accessing existing storage infrastructure is significantly less expensive than the advanced technologies.

12.5 SUMMARY

Specific actions can be taken that can enhance the value of wind energy on a power system. In general, such actions increase the economic efficiency of the power system, but may have increased importance due to the special characteristics of wind generation. Enhancements are broadly aimed at either reducing the amount of reserves necessary to maintain system reliability, or reducing the cost of providing such reserves. Both institutional and technological approaches exist to meet these challenges.

Institutional changes that enhance the value of wind energy include:

- Reducing the length of operating periods
- Increasing the frequency of intra-operating period trades
- Instituting market auctions
- Automating auctions and trade brokering
- More efficient utilization of inter-balancing area transmission capability
- Aggregating balancing requirements across balancing areas
- Consolidating multiple balancing areas together.

Technological changes to enhance wind value include:

- Improving wind generation forecasting accuracy
- Expanding transmission capability between areas experiencing congestion
- Installing phase-shifting transformers to encourage better physical allocation of power flows
- Accessing storage capability of demand-side devices
- Adding storage capability.

It should be clear from these extensive lists that there exists considerable opportunity to enhance the value of wind generation on today's power grid.

REFERENCE

Northern Tier Transmission Group (NTTG). (2008). *ACE diversity interchange evaluation*. February.

Chapter thirteen

Review of Selected Wind Integration Studies

The secret to creativity is knowing how to hide your sources.

Albert Einstein, Physicist, 1879-1955

There is a growing body of wind integration studies in North America and Europe over the past 10 years that have been conducted for a variety of purposes. Some of the purposes include:

- Determining feasibility of incorporating large amounts of wind generation
- Isolating wind integration costs
- Evaluating competing wind resource portfolios
- Determining transmission facilities needed to meet targeted levels of renewable energy production
- Estimating emissions reductions due to added wind generation
- Determining levels of wind contribution to peak demand.

The potential complexity of undertaking such studies should be clear from the foregoing chapters. It is understandable that the sophistication of wind integration studies has grown over time, and can be expected to continue to grow well into the future (Ela et al., 2009). Comparisons among studies tend to be difficult to make due to the varied nature of study purposes, the varying methodologies employed, and the range of markets and power system sizes considered. This chapter provides a summary of some of the notable studies, their purposes, methodology, and key findings. Readers are referred to these studies as representative results and methods used to date.

The studies considered here are prospective in nature—looking forward into a future in which wind has been added to an existing system. Historical loads are used for the most part, scaled up to represent demand in the future, but historical in order to be matched with wind generation for observed behavior of the wind. Most but not all of the studies represent a relatively straightforward extrapolation of present power systems,

Valuing Wind Generation on Integrated Power Systems. DOI: 10.1016/B978-0-8155-2047-4.10013-4

TABLE 13.1 Wind Integration Costs from a Sampling of Wind Integration Cost Studies

Date	Study	Wind Capacity Penetration (%)	Cost ($/MWh) Regulation	Load following	Unit commit.	Gas supply	Total
2003	Xcel-UWIG	3.5	0	0.41	1.44	NA	1.85
2003	We Energies	29	1.02	0.15	1.75	NA	2.92
2004	Xcel-MNDOC	15	0.23	NA	4.37	NA	4.60
2005	PacifiCorp-2004	11	0	1.48	3.16	NA	4.64
2006	Calif. (multi-year)*	4	0.45	Trace	Trace	NA	0.45
2006	Xcel-PSCo	15	0.2	NA	3.32	1.45	4.97
2006	MN-MISO†	36	NA	NA	NA	NA	4.41
2007	Puget Sound Energy	12	NA	NA	NA	NA	6.94
2007	Arizona Pub. Service	15	0.37	2.65	1.06	NA	4.08
2007	Avista Utilities‡	30	1.43	4.40	3.00	NA	8.84
2007	Idaho Power	20	NA	NA	NA	NA	7.92
2007	PacifiCorp-2007	18	NA	1.10	4.00	NA	5.10
2008	Xcel-PSCo	20	NA	NA	NA	NA	8.56

*Regulation costs represent 3-year average.
†Highest over 3-year evaluation period.
‡Unit commitment includes cost of wind forecast error.
Source: DeCesaro & Porter (2009).*

without taking into consideration market or technological changes that could optimize the value of wind generation as described in the previous chapter.

Keeping in mind the caveats over comparing studies across power systems and methodologies, wind integration costs found by a number of studies are shown in Table 13.1 (DeCesaro & Porter, 2009). These studies have found wind integration costs for penetration levels up to 36%[1] represent a relatively modest fraction of the value of the energy produced—in the range of roughly 5–15%.

[1] Wind nameplate capacity as a fraction of peak power system load.

13.1 SAMPLING OF STUDIES

Below are the results of some of the more recent studies that roughly represent the state of the art in wind integration studies. Given all the challenges of conducting studies described previously, each of these studies has unique strengths and weaknesses. Given the newness of undertaking these studies and their complexity, there is an evolution in which later studies by a given entity tend to better meet the challenges and improve the accuracy of results. Early studies by some entities are severely flawed and the results need to be examined carefully. Some of the more significant shortcomings of lesser wind integration studies involve some or all of the following characteristics:

- Failure to incorporate the joint variability and uncertainty of load and wind together
- Overstated correlation among wind projects modeled
- Withholding reserves for inter-hour balancing from model dispatch
- Assuming no market transactions available for system balancing
- Assuming all system balancing occurs through market transactions.

Comparing results with the better studies is an important part of validating studies. For example, a new study that shows reserve requirements or integration costs at the extreme of what the better studies have found indicates a need for closer inspection to understand whether there are compelling reasons to believe that the system under study merits a finding significantly different than those of these studies.

13.1.1 2006 Minnesota Wind Integration Study

STUDY PURPOSE AND BACKGROUND

The Minnesota Legislature acted in May 2005 to require Minnesota utilities to perform a wind integration study to assess the reliability and cost effects of matching 20% of retail sales with wind generation (EnerNex, 2006). In July 2005, the Minnesota Public Utilities Commission ordered all Minnesota utilities to participate in the study, that the study work should be conducted by an independent firm under contract, and that the findings of the studies be incorporated into Minnesota utility resource planning and renewable energy objectives reports. Key issues to be addressed by the study included:

- Assessing balancing reserve requirements to maintain reliable system operation
- Determining a 'capacity value' of wind generators
- Importance of wind generation uncertainty
- Determining system impact costs
- Assessing effects of energy markets and market structure on integrating wind generation.

Study objectives for the study were defined as:

- Evaluate the impacts on reliability and costs associated with increasing wind capacity to 15%, 20%, and 25% of Minnesota retail electric energy sales by 2020

- Identify and develop options to manage the impacts of the wind resources
- Build upon prior wind integration studies and related technical work
- Coordinate with recent and current regional power system study work
- Produce meaningful, broadly supported results through a technically rigorous study process.

One of the most significant aspects of the Minnesota Wind Study was the institution of a broad and diverse technical review committee made up of representatives from Minnesota electric utility companies, stakeholder groups, National Laboratory scientists, and state agencies. The study was competitively bid, with EnerNex Corporation winning the competition to conduct the study. The study was begun in December 2005 and completed in November 2006.

STUDY METHODOLOGY

As suggested previously, one of the most challenging aspects of wind integration studies is marshalling credible wind generation data to represent the anticipated fleet of wind generators to be installed. WindLogics, chosen to develop wind data, employed the National Center for Environmental Prediction (NCEP) MM5 mesoscale numerical weather prediction model to produce wind speed data for the study. A grid scale of 4 km was used as the finest level of detail in the MM5 runs with wind speeds reported out every 5 minutes. Wind speed validation was performed against a meteorological tower from which measurements were available at 50, 60, and 70 meters above ground.

Wind speeds were taken from the MM5 model at 152 grid points. Wind generation levels were derived by applying the wind speed sequences to a wind-turbine power curve to represent 40 MW of nameplate wind capability at the chosen grid point. Vestas V82 1.65-MW turbine power curves were used to develop the generation data from using modeled wind speeds from 80 meters above ground.

Power system dispatch was conducted using the PROMOD model from New Energy Associates. PROMOD was used to determine reserve requirements and incremental system costs. The model was also used to determine the amount of peak demand that could be depended on from the wind generators using effective load-carrying capability (ELCC) methodology based on loss of load probability (LOLP). Working with the model vendor, analysts were able to examine the benefits of time-dependent reserve requirements.

KEY FINDINGS

An important product of the Minnesota study was a map of important wind resource areas to be developed, along with the other key findings of the study. Among the key findings were:

- Wind generation levels of 15%, 20%, and 25% can be reliably accommodated in the Minnesota power system.

- Wind integration costs were $4.41 per MWh at the 25% wind penetration (by energy) level.
- Capacity values of wind generation ranged from 5% to 20% depending on penetration levels and historical wind conditions. Table 13.2 shows reserve requirements for base case and wind penetration alternatives.
- Results of the Minnesota Wind Integration Study suggest relatively modest incremental reserve requirements and wind integration costs. One of the reasons costs and reserve requirements are low may be due to the large geographic region encompassed by the analysis and participation in the large MSIO day-ahead and real-time trading markets.

13.1.2 2005 NYSERDA Wind Study

STUDY PURPOSE AND BACKGROUND

The New York State Energy Research and Development Authority (NYSERDA) and the New York Independent System Operator (NYISO) commissioned a joint study to assist NYISO in assessing the reliability impacts of wind generation on the power system (GE, 2005a, b) representative of renewable portfolio standard targets. Phase I of the study targeted reliability issues and was completed in early 2004. Phase II presented a more detailed analysis, including an economic analysis, and was completed in March 2005.

STUDY METHODOLOGY

Phase I of the NYSERDA study examined the effects of 10,026 MW of wind on the NYISO power system, representing 30% of peak load. Key

TABLE 13.2 Reserve Requirements Resulting from the Minnesota Wind Integration Study for the Base Case 2020 Loads and Alternative Wind Penetration Levels (as fraction of retail sales matched by wind generation)*

	Base		15% Wind		20% Wind		25% Wind	
Reserve Category	MW	%	MW	%	MW	%	MW	%
Regulating	137	0.65	149	0.71	153	0.73	157	0.75
Spinning	330	1.57	330	1.57	330	1.57	330	1.57
Non-spin	330	1.57	330	1.57	330	1.57	330	1.57
Load following	100	0.48	110	0.52	114	0.54	124	0.59
Operating reserve margin	152	0.73	310	1.48	408	1.94	538	2.56
Total operating reserves	1049	5.00	1229	5.86	1335	6.36	1479	7.05

Assumes 2020 MN balancing authority peak load of 20,984 MW. Requirements for load following and reserve margin based on two standard deviations of the 5-minute variability and next hour forecast error respectively.

components of the phase I study were a review of current experience with wind worldwide and a 'fatal flaw' powerflow (i.e. transmission capability) analysis. Wind site selection and hourly wind generation data were provided by AWS Scientific, Inc.

Phase II analyzed in greater detail the effects of 3300 MW of wind capability, representing 10% of peak load. Wind data were provided by AWS Truewind derived from a combination of mesoscale and microscale numerical weather models. One-minute wind data were derived from trends in hourly data based on data taken from a 105-MW wind project in Iowa. Minute-by-minute deviations were scaled to represent the appropriate smoothing effects for modeled projects. Synthetic wind forecasts were used for establishing unit commitment in the day-ahead timeframe.

Power system dispatch was modeled using General Electric's MAPS economic dispatch model. Phase II capacity value analysis was done using General Electric's stochastic MARS model using ELCC methodology based on loss of load probability (LOLP).

KEY FINDINGS

Phase I of the study concluded that there were no 'fatal flaws' in integrating 10,026 MW of wind in a system of approximately 33,000 MW of peak load. Among the key findings of the phase I report were:

- Certain interconnection requirements should be met (low-voltage ride-through, voltage regulation, project metering and monitoring), and power curtailment should be entertained for projects larger than 5 or 10 MW.
- New York State should develop a centralized wind power forecasting capability.
- No integration issues are expected with penetrations up to at least 10% of peak load (3300 MW).
- Some wind generation limits may be necessary under light load conditions with 10,026 MW of wind.

The phase II study findings included:

- Confirmation that 10% wind penetration can be accommodated with relatively minor operational adjustments.
- Dispatch model found a $1.80/MWh average reduction in wholesale prices due to the presence of wind generation.
- Wind energy was worth an average of $48/MWh in reduced operating costs when wind forecasts were included in the day-ahead unit commitment.
- Leaving forecast wind energy levels out of the day-ahead unit commitment reduced the operational benefits by $10/MWh (effective value of day-ahead forecast) for each megawatt-hour of wind generation.
- Wind energy reduced generation and emissions from other generation, with 65% of the wind energy displacing natural gas generation, 15% coal, 10% oil, and 10% imports.

- Capacity value of onshore wind was found to be roughly 10%; the one offshore site examined appeared to have a capacity value of about 36%.
- Analysis showed a 6% increase in hourly variability (standard deviation) and a 3% increase in 5-minute variability.
- Regulating reserve requirements were increased by about 14% to maintain the historical reliability levels (above minimum standard), and would probably continue to exceed standard without adding reserves.
- Market rules and incentives should be structured to provide incentives for improved forecasts.
- A need to provide incentives to encourage wind projects to reduce output during light load periods when wind generation would otherwise be excessive.

13.1.3 California Energy Commission 2007 IAP Final Report

STUDY PURPOSE AND BACKGROUND

The California Energy Commission organized an Intermittency Analysis Project (IAP) to consider the effects of high levels of renewable generation on California's power grid (Porter, 2007). IAP issued a series of seven reports with the help of: BEW Engineering for reporting on wind-turbine technology developments; Exeter Associates, Inc. to report on international experience; AWS TrueWind for wind generation and forecast data; Davis Power Consultants to develop renewable resource mixes, load flow analysis, and transmission build-out scenarios; and General Electric Energy Consulting for statistical analysis and production cost modeling.

Project objectives included:

- Determining transmission options to meet state renewable energy targets (20% by 2010, 33% by 2020)
- Evaluating effects of large-scale renewable energy development on transmission reliability and resource dispatch
- How the power system will need to change to accommodate the studied renewable resource development
- Developing analytical capability to evaluate renewable resources on a par with conventional generation.

STUDY METHODOLOGY

The IAP team modeled transmission load flows, developed statistical analyses of load and wind generation, performed production cost model studies, and examined the statewide power system impacts of renewable generation on the power system. Four scenarios were examined:

- 2006 base—existing levels of wind (2100 MW), with renewable energy supplying 5% of energy in the independent system operator (ISO)
- 2010T—assumed adding 3000 MW of wind generation in the Tehachapi area, for a total of 7500 MW wind in the ISO (18% renewable energy case)

- 2010X—included 12,500 MW of wind in the ISO (29% renewable case)
- 2020—included 12,700 MW of wind in the ISO (31% renewable case).

Wind site selection and generation data development were done by AWS TrueWind. Hourly wind generation sequences were produced for historical years 2002–4 using the mesoscale weather model on 8-km grids. Wind speed and direction data were applied to wind-turbine power curves. Several power curves were employed, chosen on the basis of specific site conditions (wind class). Power curve generation levels were reduced an average of 14% for estimates of losses such as electrical system, wake effects, turbine availability, turbulence, and blade soiling. One-minute wind data were simulated for selected time periods based on the statistical characteristics of a sample data set from an Iowa wind project. Simulated wind generation series were validated against existing wind projects at Altamont Pass and Tehachapi Pass.

Power flow modeling determined transmission reliability under the scenarios examined. General Electric's MAPS economic dispatch model processed hourly load and generation data to determine impacts to system operations including operational impacts, reductions in emissions, and power production costs. Statistical analysis of selected historical periods of concern (low loads and high wind, or times of high wind variability) using 1-minute data provided more detailed analysis where needed.

KEY FINDINGS

The principal conclusion of the study is that the levels of renewable energy examined are feasible assuming implementation of certain infrastructure, technology, and policy changes. More specifically, study findings included:

- Additional transmission facilities are needed.
- Renewable generation displaces generation with higher marginal costs, primarily combined-cycle natural gas generators.
- Under some low-load, high-wind conditions it may be necessary to limit wind generation.
- Wind forecasting in both day-ahead (unit commitment) and hour-ahead timeframes have high value and improvements in forecasting should be incentivized.
- System ramping capability requirements increase by 130 MW/hour due to renewable generation.
- Load-following balancing requirements increase by 3–7%.
- Regulation balancing requirements also increase by 3–7%.
- Reductions in power system emissions represent a major benefit of wind generation with reduction in NO_x of 117 pounds/MWh, SO_x of 46 pounds/MWh, and CO_2 of 810 pounds/MWh of wind generation.

13.1.4 Eastern Wind Integration and Transmission Study (EWITS)

STUDY PURPOSE AND BACKGROUND

The 2010 Eastern Wind Integration and Transmission Study (EnerNex, 2010) was commissioned by the National Renewable Energy Laboratory (NREL) to provide a more detailed power system analysis to follow the US Department of Energy 2008 study of 20% of electric energy coming from wind power by 2030 (US DOE, 2008). Above all, the study sought to provide an objective technical analysis of a high wind penetration future to help power system planners prepare for such a future. A large and diverse technical review committee provided oversight and review of the work, and included representatives from the affected ISOs, utilities, consultants, government energy agencies, regulatory bodies, national laboratories, the Utility Wind Integration Group (UWIG), and universities. One of three concurrent studies of large-scale wind integration over a large geographic footprint, the EWITS footprint is roughly the eastern third of the USA and Canada (excluding Texas and Quebec). The two sister studies are the Western Wind and Solar Integration Study (WWSIS) and the European Wind Integration Study (EWIS).

STUDY METHODOLOGY

Three of the four scenarios analyzed represented different combinations of wind sites that reached an overall wind penetration of 20%. The fourth scenario included wind at a 30% penetration level (energy basis). A fifth reference scenario was also designed to represent the current level of wind development and some near-term development as suggested by grid operator interconnection queues. Details of the placement of wind in the EWITS scenarios are presented in Table 13.3, but the scenarios in brief were:

- **Scenario 1.** 20% penetration—high capacity factor, onshore: includes high-quality wind resources in the Great Plains, with other development in the eastern USA where good wind resources exist.
- **Scenario 2.** 20% penetration—hybrid with offshore: some wind generation in the Great Plains is moved east. Some East Coast offshore development is included.
- **Scenario 3.** 20% penetration—local with aggressive offshore: more wind generation is moved east toward load centers, necessitating broader use of offshore resources. The offshore wind assumptions represent an uppermost limit of what could be developed by 2024 under an aggressive technology-push scenario.
- **Scenario 4.** 30% penetration—aggressive on- and offshore: meeting the 30% energy penetration level uses a substantial amount of the higher-quality wind resource in the NREL database. A large amount of offshore generation is needed to reach the target energy level.

- **Reference scenario.** 6% penetration—approximates current state of wind development plus some expected level of near-term wind development in compliance with existing renewable portfolio standards.

An earlier study, the Eastern Wind Data Study, developed wind generation time series for the historical years 2004–6 that was used in the EWITS. Wind speed data were produced by AWS TrueWind with a 2-km grid mesoscale model reporting wind speeds at 10-minute intervals.

Conversion from wind speeds for each 2-km model grid began with an adjustment to account for differences between simulated wind speeds and available historical meteorological tower measurements. Simulated wind speed data for grids representing the geographic vicinity of each wind project site were summed as weighted averages to depict the number of wind turbines expected in each grid. These wind speeds were reduced by a wind speed loss factor representing wake losses assumed to be in the range of 4% (w_{min}) to 9% (w_{max}) depending on wind direction according to the formula:

$$w = (w_{max} - w_{min})\sin^2(\theta - \theta_{max})$$

where w is the percentage of lost wind energy and θ is the direction of the wind and θ_{max} is the direction of the most prevalent wind for the particular

TABLE 13.3 Wind Additions (MW) in Each of the EWITS Scenarios

Zone	Scenario 1	Scenario 2	Scenario 3	Scenario 4
MISO West	59,260	39,953	23,656	59,260
MISO Central	12,193	11,380	11,380	12,193
MISO East	9091	6456	4284	9091
MAPP USA	13,809	11,655	6935	14,047
SPP North	48,243	40,394	24,961	50,326
SPP Central	44,055	46,272	25,997	44,705
PJM	22,669	33,192	78,736	93,736
TVA	1247	1247	1247	1247
SERC	1009	5009	5009	5009
NYISO	7742	16,507	23,167	23,167
ISO-NE	4291	13,837	24,927	24,927
Entergy	0	0	0	0
IESOa	0	0	0	0
MAPP Canada	0	0	0	0
Full Study System	223,609	225,902	230,299	337,708

site. A further adjustment is taken for higher or lower wind turbulence than the expected value. Resulting wind speeds are then averaged over a 3-hour time window to represent the effects of wind turbines spread over large geographic areas and to reproduce the expected variability. A final additional loss is taken to represent wind turbine outages. Total losses applied to the data averaged roughly 15–17%.

Wind generation and load data for the historical period were input to the PROMOD IV chronological economic dispatch model. Economics-based capacity expansion methods were used to determine levels of traditional generation required to meet load assuming wind contributed 20% of nameplate capacity to meeting peak demand. Next, wind generation was added to the model. The value of transmission constraints was estimated by running the models with existing transmission path limits, and with unlimited transmission. Analysts also assumed area-wide markets for moving generation to load in an economic manner.

The capacity value of wind in the scenarios was determined through ELCC methods on a loss of load expectation basis, using the chronological economic dispatch model. Figure 13.1 shows a process diagram of the steps undertaken.

KEY FINDINGS

Study results showed that wind generation could supply 20–30% of the electrical energy requirements of the Eastern Interconnection under the study assumptions, including significant expansion of the transmission system. Other key findings of the study include:

- New transmission will be required for all the future wind scenarios in the Eastern Interconnection, including the Reference Case. Planning for this transmission is imperative due to the long lead times involved with adding transmission capacity compared with adding wind generation.
- Significant curtailment (limiting generation levels) of wind generation would be required for all the 20% scenarios without transmission expansion.
- Wind integration costs associated with large amounts of wind generation are manageable with large regional operating pools and significant market, tariff, and operational changes.
- Minimizing transmission congestion reduces the impacts of the variability of the wind, reduces wind integration costs, increases reliability of the electrical grid, and helps make more efficient use of the available generation resources. Although costs for aggressive expansions of the existing grid are significant, they make up a relatively small portion of the total annualized costs in any of the scenarios studied.
- Balancing reserve requirements increased from 5% to 5.5% of wind nameplate generation.
- Integration costs varied by historical year and by scenario, but in all cases were less than 10% of the busbar cost of wind generation.

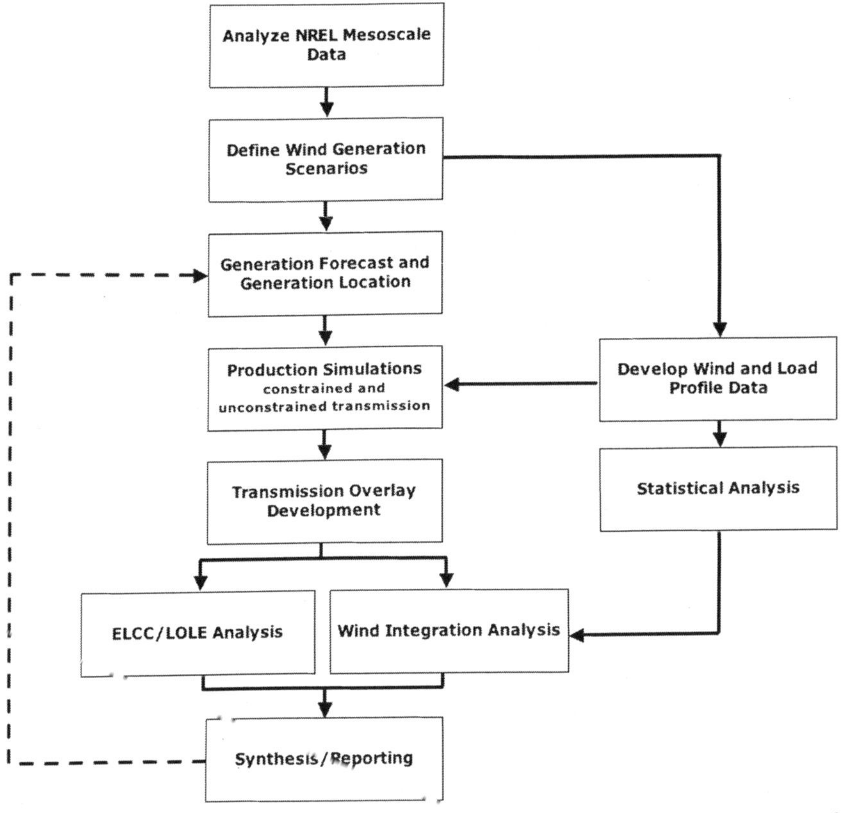

FIGURE 13.1 Process diagram of the EWITS study, Note that LOLE is the loss of load expectation and ELCC is effective load-carrying capability. *Adapted from EnerNex (2010).*

- Capacity value as determined by ELCC methods ranged over scenarios from 16% to 33%, strongly dependent on available transmission and fraction of wind coming from offshore wind projects.

13.1.5 Western Wind and Solar Integration Study (WWSIS)

STUDY PURPOSE AND BACKGROUND

The Western Wind and Solar Integration Study (WWSIS) is the western US counterpart to the Eastern Wind Integration and Transmission (EWITS) study commissioned by the US National Renewable Energy Laboratory (Piwko et al., 2010) to provide additional detail and confirmation for the US Department of Energy study (US DOE, 2008), "20% Wind Energy by 2030". Because the earlier study assumed 25% wind penetration in the Western Electricity Coordinating Council (WECC) region, the WWSIS investigated the operational effects of up to 35% of energy coming from renewable resources (wind and solar) in the WestConnect utilities. Figure 13.2 shows the WestConnect geographic footprint.

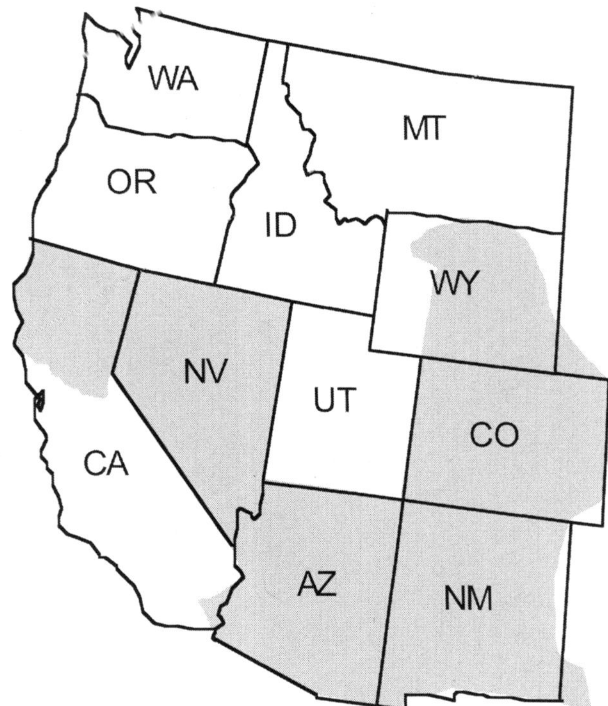

FIGURE 13.2 Geographic region served by WestConnect utilities in the western USA shown in gray.

Funded by the US Department of Energy, the WWSIS was managed by the National Renewable Energy Laboratory that assembled a project team including 3Tier (provider of wind data and solar forecasting services), State University of New York (solar radiation data set), Exeter Associates (data collection), Northern Arizona University (wind validation, hydro), NREL (wind validation, PV and CSP datasets), and General Electric (scenario development, analysis). In addition, a technical review committee including WestConnect utility representatives, utility organizations, industry, and technical experts met to review study results and progress. A broader stakeholder group to which the public was invited met four times to provide additional input and broadly vet study assumptions and results.

STUDY METHODOLOGY

The WWSIS examined the operational power system impacts associated with a large penetration of renewable energy assumed in service in the year 2017. Historical load and weather patterns (i.e. wind and solar performance) were taken from 2004, 2005, and 2006. Wind penetration rates up to 30% of energy demand were considered. Due to the abundance of solar resource in the region, solar energy was included.

While the study focused on a subset of the WECC region, assumptions about the amount of renewable resources in the surrounding markets were necessary. Some 75 GW of wind generation sites were necessary for study scenarios. Ultimately the datasets provided by 3Tier and SUNY included 960 GW of wind generation, 15 GW of solar PV, and over 200 MW of concentrating solar power (CSP) facilities. Wind data were produced using mesoscale modeling techniques at 10-minute intervals, aggregated into 1-hour periods for modeling purposes. Forecasts of wind and solar, along with suitable forecast errors, were an integral part of the analysis.

Combinations of wind and solar penetration levels in WestConnect and in the remaining WECC region were examined, along with a range of sensitivities to fuel costs, operating reserve levels, unit-commitment strategies, storage alternatives (including rechargeable hybrid electric vehicles), and balancing area size. Table 13.4 shows the renewable resource penetration levels in the study footprint and surrounding region. Annual WestConnect region loads were 286,000 GWh.

In addition to the renewable penetration rate assumptions, three scenarios examining different geographic locations of renewable resources were also tested. Each of the three scenarios represented a distinct strategy for resource development:

- **In-area scenario** assumed that renewable resource target levels were met by each state individually, resulting in a broader distribution of resources than the other scenarios.
- **Mega-project scenario** assumed that the best resources available in the footprint would be developed to meet the penetration targets. This scenario resulted in much of the wind generation sites locating in wind-rich Wyoming with new transmission facilities to move the energy to load.
- **Local priority scenario** allowed the best wind and solar sites within the footprint to be developed, but accorded a 10% capital cost

TABLE 13.4 Wind and Solar Penetration Levels as a Fraction of Total Electric Energy Consumed in the WWSIS Scenarios

	Penetration in Study Footpring			Surrounding Area	
Case Name	Wind + Solar	Wind	Solar	Wind	Solar
10%	11%	10%	1%	10%	1%
20%	23%	20%	3%	10%	1%
10/20%	23%	20%	3%	20%	3%
30%	35%	30%	5%	20%	3%

advantage to resources within each state to meet state targets. The scenario resulted in a geographic distribution roughly midway between the other two scenarios, necessitating more cross-state transmission facilities than the in-area scenario, but less than the mega-project scenario.

The resulting geographic distributions of the renewable generation in each of the scenarios, along with the needed interstate transmission paths, is illustrated in Figure 13.3.

Power system dispatch was simulated using the General Electric MAPS chronological economic dispatch model simulating power system operations in hourly time increments. The WECC region was modeled as a set of 106 zones, each represented by its own loads, generators, and transmission interconnection with neighboring zones. Model simulations assessed costs and benefits of the scenarios and quantified:

- Available maneuverable generation, including up- and down-ramping capability by hour
- Effects of day-ahead wind forecasts in unit-commitment decisions
- Changes in conventional generation operations
- Emissions (NO_x, SO_x, and CO_2) due to renewable generation
- Transmission path loading
- Value of energy storage.

Grid performance at the sub-hourly level was analyzed separately by evaluating specific challenging events (e.g. loads and high wind, or rapid load/wind ramp events) using 1-minute data. A separate stochastic analysis using the General Electric MARS program evaluated load service reliability (loss of load expectation) and determined the capacity value of wind using ELCC methods.

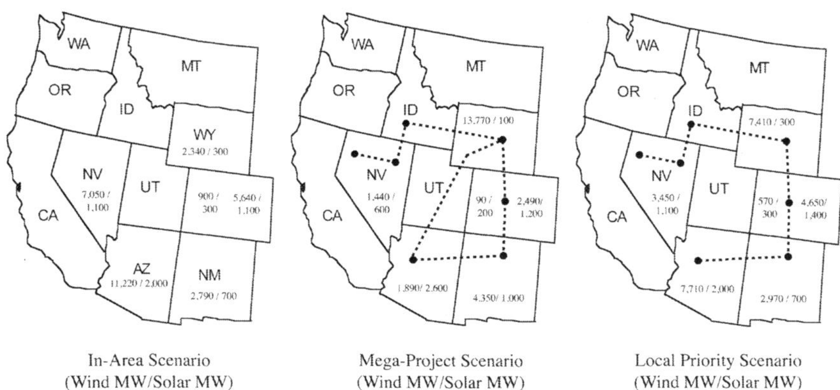

In-Area Scenario (Wind MW/Solar MW)

Mega-Project Scenario (Wind MW/Solar MW)

Local Priority Scenario (Wind MW/Solar MW)

FIGURE 13.3 Renewable energy geographic distributions resulting from the three build strategies. In each strategy, 45,400 MW of wind and 7500 MW of solar were built into the balance of the WECC.

KEY FINDINGS

Results of the WWSIS confirmed that integrating 30% wind energy and 5% solar energy is technically feasible. Other findings include:

- Operational challenges are greatest in lower-load months (e.g. spring), but ensuring system reliability during those times would require extensive balancing area cooperation or consolidation.
- Incorporating state-of-the-art day-ahead wind and solar forecasts for the purpose of unit-commitment decisions is essential, reducing operating costs by $5 billion across the WECC annually. Perfect forecasting reduced costs another $500 million annually.
- Few significant operational concerns manifest up to about 20% wind penetration.
- The need for flexible generation or load becomes more significant beyond about 20% wind penetration levels.
- Sub-hourly scheduling or dispatch of resources is needed at high penetration levels to keep reserve requirements within bounds.
- At 30% penetration, the presence of wind and solar generation on the system reduced balance of the system costs by about $80/MWh for each megawatt-hour of renewable energy produced. At lower levels, the value was approximately $88/MWh.
- WECC-region carbon dioxide emissions dropped 20% in the 30% scenario.
- The capacity value of wind generation ranged from 10% to 15% in the study scenarios.
- Existing storage facilities (e.g. hydro and pump storage) were utilized to a greater degree in the studies, but new storage facilities were not economically beneficial.
- Wind curtailments contributed to maintaining system reliability, but did not constitute a significant reduction in wind energy production (0.5% of wind energy curtailed in the 30% scenario) due to the infrequency of such events.
- Load participation in providing balancing services is an economic alternative to holding additional reserves or building additional power plants to provide balancing services.
- Wind and solar energy primarily displace natural-gas-fueled generation.

13.1.6 All Island Study (Ireland)

STUDY PURPOSE AND BACKGROUND

Ireland's Department of Enterprise, Trade and Investment undertook the All Island Grid Study (DETC, 2008) to provide a comprehensive assessment of the ability of the electricity transmission network on the island of Ireland (Northern Ireland and the Republic of Ireland together) to absorb large amounts of renewable electric energy. In 2005 the then Department of Communications, Marine and Natural Resources in the Republic of Ireland and the Department of Enterprise, Trade and Investment in

Northern Ireland jointly issued a preliminary consultation paper on an all-island '2020 Vision' for renewable energy. The paper solicited views on developing a joint strategy for developing renewable energy within the All-island Energy Market leading up to 2020 and beyond, so that 'consumers, North and South, continue to benefit from access to sustainable energy supplies provided at a competitive cost.'

Following the All-island Energy Market Development Framework agreed by Ministers and efforts to develop a single electricity market, concerns turned to how the electricity infrastructure might best develop to allow the efficient utilization of high levels of renewable energy. A working group was established that recommended conducting the All Island Study.

STUDY METHODOLOGY

The All Island Study consisted of six generation portfolios comprising a range of different renewable energy penetration rates and configurations. Resource portfolios were constructed through an optimization process to choose the least cost set of resources to meet target resource penetration levels and combinations of resource types. Portfolio 1 resulted in 16% renewable energy (2000 MW of wind), portfolios 2–4 met 27% (4000 MW wind) with renewable energy, and portfolio 5 met 42% (6000 MW of wind). A sixth portfolio targeting 59% renewable energy was not pursued as it resulted in unacceptable operations under the operational assumptions of the study. Portfolio 4 included a large new coal plant.

Costs and benefits were assessed for the specific year 2020. The work was broken down into separate 'work stream' studies involving specialist consultants with relevant expertise for each area to develop: a screening study, a resource assessment, a dispatch study, and a network study, all of which provided information for the concluding cost–benefit study. Although the work streams effectively represented different studies, the studies were closely interrelated, as shown in Figure 13.4.

Resources were chosen without respect to transmission availability in work streams 1 and 2a. Work stream 3 determined the needed transmission additions to accommodate the added resources and assess transmission costs. Fuel consumption for each of the portfolios is shown in Figure 13.5. The study assumed 1000 MW of transmission interconnection with Great Britain, 100 MW of which was reserved for spinning reserve.

KEY FINDINGS

Ireland's All Island Grid Study concluded that:

- Up to 42% of Ireland's energy can come from renewable energy in 2020 (Portfolio 5), resulting in 25% lower carbon dioxide emissions than the portfolio 1 16% renewable energy reference case.

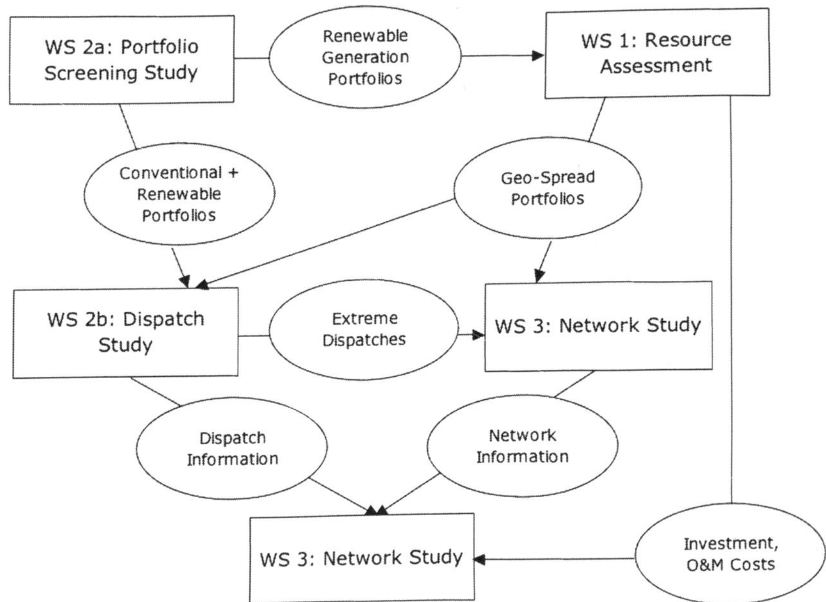

FIGURE 13.4 Process diagram for the work streams ('WS' in rectangles) and products (in ovals) for the All Island Study. *Adapted from DETC (2008).*

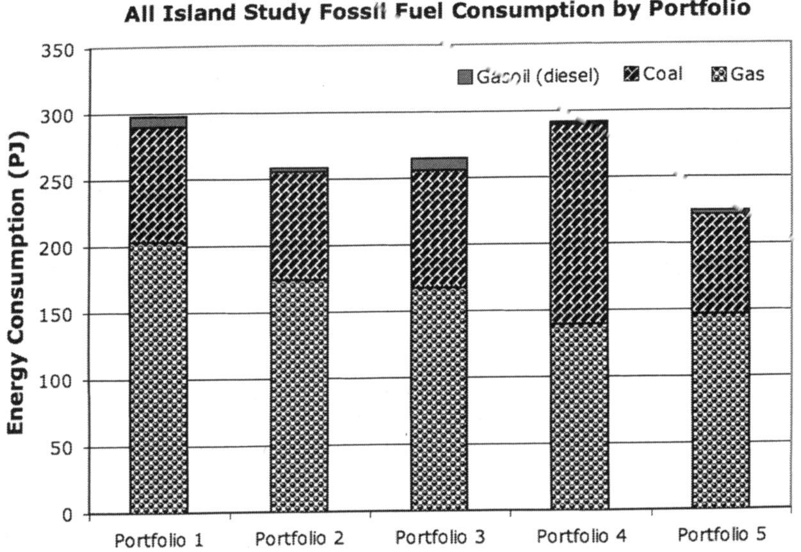

FIGURE 13.5 Breakdown of fossil-fuel consumption of All Island Study portfolios.

- The incremental costs of reducing carbon emissions by 25% in portfolio 5 over portfolio 1 were 7%, for a rate impact of about €4 per MWh.
- More moderate portfolios 2 and 3 reduced carbon emissions 10% at a cost of about €2 per MWh—portfolio 4 with additional coal resources increased carbon emissions of portfolio 1.
- Market mechanisms are needed to facilitate development of complementary (flexible) dispatchable loads or resources to maintain adequate system security.

13.2 SUMMARY

Highly sophisticated wind integration studies have repeatedly shown the feasibility of operating power systems with wind levels in the 10–20% range, with some studies suggesting feasibility for as much as double that range. Studies consistently show wind generation imposing relatively modest 'integration costs' on the balance of the power system, and that wind generation tends to displace fossil resources and reduce associated emissions. A number of studies using ELCC methods show a capacity value of wind projects in a range from about 10% to as much as 33% depending on penetration level and the amount of offshore wind resource included. The studies reported here examined systems without extensive modifications such as the addition of new generating capability or storage capacity. On the other hand, the need for expanded transmission facilities was common, as well as the need for greater regional and inter-regional cooperation among balancing areas.

REFERENCES

DeCesaro, J., & Porter, K. (2009). *Wind energy and power system operations: A review of wind integration studies to date. National Renewable Energy Laboratory Subcontractor Report NREL/SR-550-47256*. December.

Department of Enterprise, Trade and Investment (DETC). (2008). *All Island Grid Study: Study overview*. January.

Ela, E., Milligan, M., Parsons, B., Lew, D., & Corbus, D. (2009). *The evolution of wind power integration studies: Past, present, and future. IEEE Power and Engineering Society General Meeting PES '09, IEEE*. July.

EnerNex. (2006). *Final report—2006 Minnesota Wind Study*. EnerNex Corporation. November.

EnerNex. (2010). *Eastern Wind Integration and Transmission Study*. EnerNex Coproration. January.

GE Power Systems Energy Consulting. (2005a). *The effects of integrating wind power on transmission system planning, reliability, and operations. Phase I report February 2004, phase II report March, 2005*. NYSERDA.

GE Power Systems Energy Consulting. (2005b). *The effects of integrating wind power on transmission system planning, reliability, and operations: Report on phase 2 system performance evaluation. Prepared for the New York Energy and Research and Development Authority*. March 2005.

Piwko, R., Clark, K., Freeman, L., Jordan, G., & Miller, N. (2010). *Western wind and solar integration study: Draft executive summary*. National Renewable Energy Laboratory Preprint. February.

Porter, K., & , and Intermittency Analysis Project Team. (2007). *Intermittency Analysis Project: Summary of final results. California Energy Commission, PIER Research Development & Demonstration Program*. CEC-500-2007-081.

US Department of Energy (US DOE). (2008). *20% Wind energy by 2030: Increasing wind energy's contribution to US electricity supply*. July. <http://www.20percentwind.org/20percent_wind_energy_report_revOct08.pdf>

Chapter | fourteen

Considerations for High Penetration Wind Systems

Life is either a daring adventure, or nothing.
Helen Keller, American author, activist, and lecturer, 1880–1968

Today's power systems and the institutions that have grown up around them were not designed to make optimal use of more variable and less controllable wind generation. Historically, the foremost concern for power system planners and designers has been ensuring the stability and reliability during extremes of power demand. Additional generators were indicated when the maximum demand levels exceed the expected capability of the available generation. The maximum capability of the transmission system to convey power was similarly calculated under extreme demand scenarios, and rights to using the transmission were almost universally sold on the basis of the maximum power. Fundamental to these assumptions is that the generation of electric power responds to the uncontrollable demand for that power, and that the power system's capacity for meeting the peak demand is the scarce commodity that must be carefully monitored, preserved, and allocated. Wind generation is fundamentally different, and it may be useful to step back and consider the attributes of a power system designed and operated to make optimal use of wind energy.

Denmark currently leads the world in wind energy production relative to load. With 20% of electric consumption now coming from wind generators, Denmark targets 50% by 2020 and is not planning to stop there. Dedicated to eliminating fossil-fuel consumption entirely, situated at a northern latitude somewhat less conducive to solar energy applications, the country has embraced wind energy. The challenge of integrating

Valuing Wind Generation on Integrated Power Systems. DOI: 10.1016/B978-0-8155-2047-4.10014-6

so much wind energy into the power grid has become a national focus. Henrik Bindslev, Executive Director of Denmark's Risø National Laboratory explains that Denmark expects to transform the paradigm under which the power grid is designed and operated. The current system in which generators respond to the whims of demand will become a system in which the demand will respond to changes in generation. This fundamental shift in thinking, planning, and designing will be needed to accommodate a power grid in which 50% or more of the energy is supplied by the wind.

Although some wind sites exhibit capacity factors that rival those of thermal power plants, the basic variability inherent in even large and diverse wind generation fleets cannot be ignored. For example, Figures 14.1 and 14.2 show 2009 production levels of wind connected to the Bonneville Power Administration (≈2300 MW of installed capacity by end of 2009) and Denmark (≈3500 MW capacity). It is clear from these figures that at least some systems with relatively large amounts of wind generation, spread over significant geographic regions, can expect significant variability in production. Figure 14.3 illustrates the same data organized to show percentage of hours that given generation levels were exceeded in 2009. Note that both Denmark and the Bonneville Power Administration (BPA) experienced significant growth in wind generators from the beginning of the year to the end.

Wind generation levels for some sites, especially offshore wind sites, may exhibit significantly less variability than that experienced by the BPA and Danish systems. However, it is likely that, for most systems, the kind of variability shown in these figures is not likely to go away entirely. As of 2009, approximately 20% of Danish electric power was supplied by wind. With plans to meet 50% of electric energy consumption with wind energy

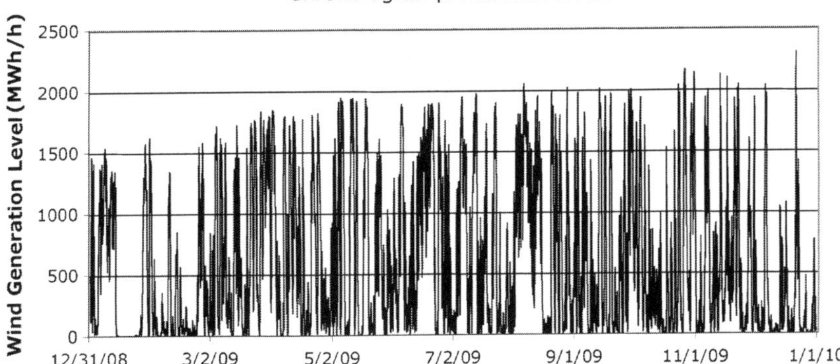

FIGURE 14.1 Hourly chronological wind generation on the Bonneville Power Administration grid in 2009. Wind nameplate capacity was 1633 MW at the beginning of the year and 2357 MW by the end of the year.

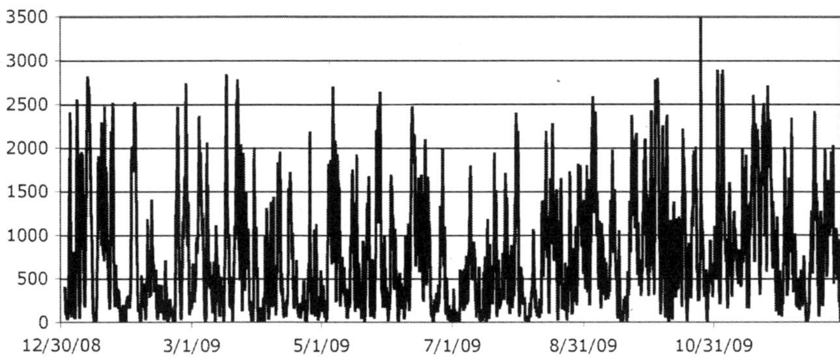

FIGURE 14.2 Danish hourly chronological wind generation in 2009. Wind nameplate capacity ended the year near 3500 MW. Note that the data here combine east and west Denmark generation. The two halves of the Danish system were not interconnected with one another in 2009.

by 2020, Danish planners are uniquely focused on how to deal with the variability of wind generation output. The two biggest challenges of the Danish system are what to do with generation during periods of low intrinsic demand, and where fossil-free power will come from when the wind does not blow.

Those challenges can be met in several ways. District heating systems provide space and water heating services for 60% of Danish households. Wind generation during light load hours can be stored as thermal energy for use in the district heating system by incorporating electric resistance heat boilers into the system. Many district heating systems already

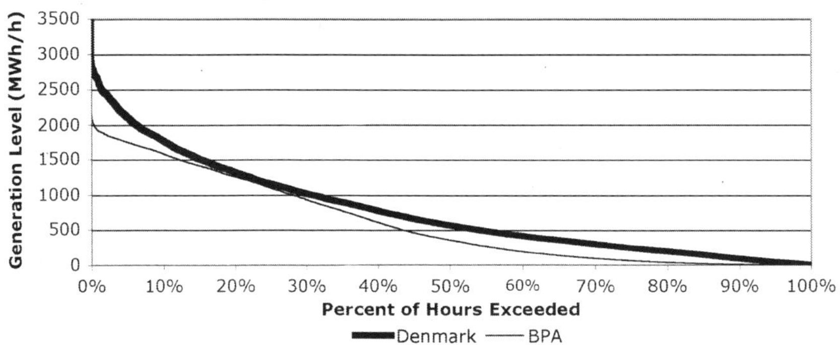

FIGURE 14.3 The distribution of generating levels for Bonneville Power Administration (BPA) and Danish hourly wind generation for 2009. The chart shows, for example, that wind generation exceeded 500 MW approximately 43% of all hours in 2009 for BPA and 54% in Denmark.

FIGURE 14.4 District heating thermal storage facilities at the Avedøre 2 combined heat and power plant near Copenhagen, Denmark.

incorporate large thermal energy storage capability. Figure 14.4 shows the thermal storage tanks from the Avedøre 2 multi-fuel (straw, wood, oil, and natural gas) power plant near Copenhagen.

The remainder of this chapter will contemplate an efficient power system in which at least half the energy produced comes from wind generation.

14.1 MARKET ORGANIZATION

A power system designed to make the highest possible use of wind energy will access flexible generation and loads across the widest possible geographic region. In the USA, existing independent system operators (ISOs) will expand to absorb neighboring balancing areas not currently served by an ISO and new ISOs will form from currently separate balancing areas. Trade barriers are minimized through automated auctions that take place on day-ahead, hour-ahead, and 5-minute bases. Models of transmission congestions are updated and operate with such rapidity that gate closure times for hour-ahead trading occur no more than 30 minutes prior to the operating period. Operating periods themselves are no longer than 30 minutes long.

Five-minute response balancing service auctions minimize the cost of accommodating wind variability and uncertainty by being well attended by load control aggregators. These providers of balancing services will include all thermal storage electric energy users such as residential space conditioning, refrigeration, and water-heating loads. Commercial buildings, municipal water pumps, and district heating systems all bid into the balancing service markets to absorb low-cost and relatively low-value wind generation during those occasions when consumption is otherwise low and the wind is strong.

Aggregation of balancing areas will occur such that the western third of the USA, currently served by dozens of balancing areas, develops one or a few liquid markets that vastly improves the ability to trade across balancing area borders. The practice of bilateral reservations of transmission services is replaced by a combination of markets and bilateral 'contracts for differences' (see Chapter 12). The need for new transmission and management of congestion within the expanded balancing area is managed by the single entity in coordination with appropriate stakeholders.

In the grid of the future, reduction of fossil-fuel generation will become a goal in North America as it is in Europe. Carbon taxes, limits on allowable carbon emissions, or rules that specifically favor renewable generation will be imposed that cause carbon-intensive resources such as coal to be more readily displaced by lower-carbon resources.

14.2 ENERGY STORAGE

Creating the power grid of Dr Bindslev's vision in which demand responds to supply primarily means accessing the ability of many electric power end uses to store energy for some period of time. Much discussion and thought is currently centered on accessing battery storage capability in the fleet of new electric vehicles that is anticipated to materialize over the next few decades. Although this is an important and exciting new source of power system flexibility, its application is likely somewhat limited to relatively short-term fluctuations. As suggested in Chapter 12, even the widespread use of electric vehicles in the USA is unlikely to add more energy storage capability than one of the larger US hydroelectric storage reservoirs.

Electric vehicle batteries are not the only storage opportunity, however. Denmark serves 60% of buildings with district heating systems that convey 'waste heat' from thermal power plants (some burning straw or wood pellets) for water heating and space conditioning purposes. To accommodate the varying demand for heat in these systems, it is not uncommon for district heating systems to employ significant thermal storage facilities. Figure 14.4 shows the thermal storage facilities at the Avedøre 2 combined heat and power facility near Copenhagen. The two tanks store approximately 22,000 cubic meters of water each at temperatures just over 100°C, with a total of about 2500 MWh of thermal energy storage—equivalent to roughly 100,000 electric vehicles.

Avdedøre 2 does not currently have electric heating elements to absorb power when high levels of wind energy push wholesale market prices down. The main reason for this is the high electricity taxes levied in Denmark. However, the Danish government has responded, removing the electricity tax for power used to supply district heating systems in 2008. In 2010 there were at least four district heating facilities that have added electric boilers and wrapped their storage tanks with 30 cm of insulation (Energinet.dk, 2009). The district heating plant in Skagen, Denmark is one of the four such systems, having added a 10-MW boiler to complement its 4150-cubic-meter, 250-MWh storage capability. The Skagen plant operator estimates that running just 800 hours turns a profit on the electric boiler, which also receives a credit for providing down-regulation services.

Attaching electric resistance boilers to access this considerable storage capability can be a low-cost way to provide balancing services. Consider, for example, the effect of a 300-MW electric boiler on the balancing requirements for a system with significant wind generating capability. Figure 14.5 shows sample wind and load data on the Bonneville Power Administration (BPA) grid. Balancing requirements for wind and load over the same time period are shown in Figure 14.6. Figure 14.7 shows the considerable reduction in reserve deployment due to operation of the electric boiler with 5000 MWh of thermal energy storage.

It is apparent from the sample data that relatively modest thermal storage capability can have significant impacts on the need for power system flexibility, shedding light on Dr Bindslev's vision of a grid where load responds to generation. This same strategy could be combined with an energy purchase (or added baseload generation) to reduce the need for both incremental and decremental reserves. Conversely, greater

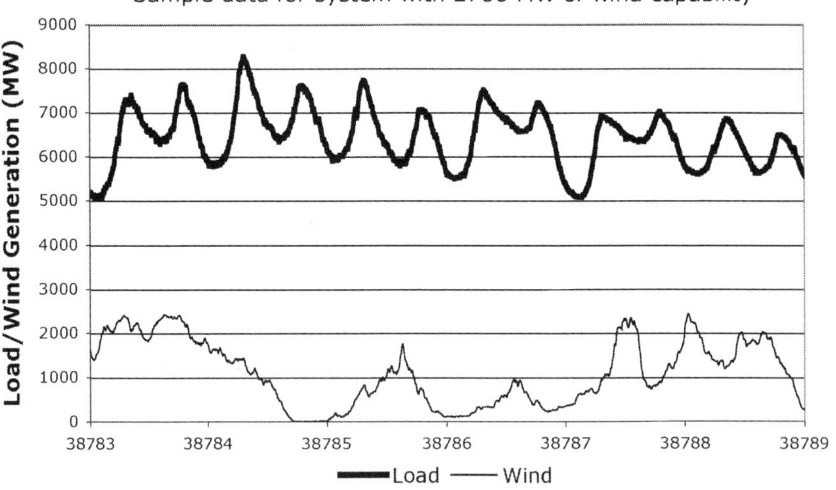

FIGURE 14.5 Sample wind and load data for the Bonneville Power Administration power grid, featuring 2780 MW of wind nameplate capacity.

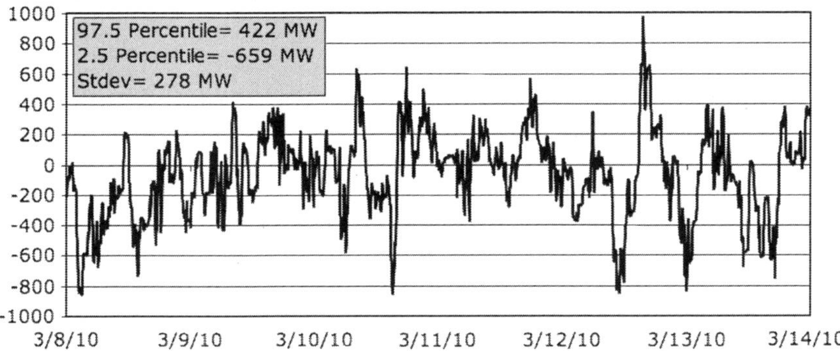

FIGURE 14.6 Balancing requirements (load net of wind) over the same time period as that used in Figure 14.5. Positive values indicate a need to supply generation from reserve units, negative values indicate a need to reduce generation or increase dispatchable demand such as energizing electric water heater loads.

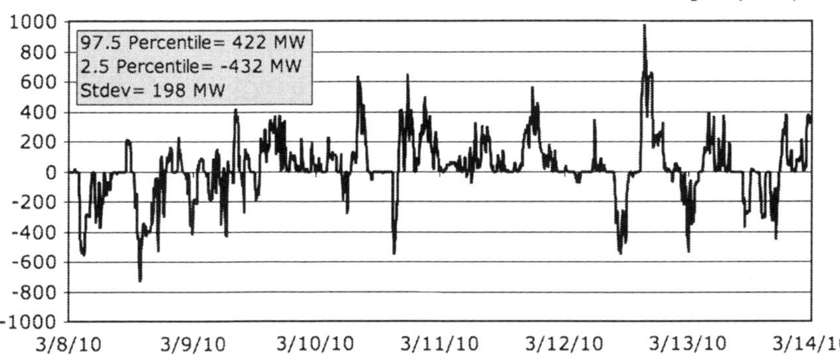

FIGURE 14.7 The effect on the balancing requirements in Figure 14.6 after the addition of a 300-MW dispatchable electric boiler load and 5000 MWh of thermal storage capacity. Balancing requirements and variability are significantly reduced.

reliance could be placed on decremental reserves by scheduling less of the expected wind generation. For example, the wind schedules could be based on an 80% probability of being exceeded instead of 50%, as is usually the case. The effect would be to make the system generally more surplus.

In North America, where district heating and cooling systems are relatively limited, other approaches may be more cost-effectively implemented. One such approach may be micro-scale combined heat and power production. Such technology does exist and might be combined with electric resistance backup to provide similar services.

More traditional load control technologies are also available such as intentional operation of space conditioning, commercial and residential water heating, refrigeration and municipal water supply pumps. The importance of these technologies is that they exist and have been financed principally on the basis of performing services other than to provide power system balancing (e.g. acceptable building temperatures, hot water, municipal water supplies, etc.). Utilizing existing infrastructure in this way is presumably vastly less expensive than the cost of all-new dedicated infrastructure such as pump storage or compressed air facilities. Many of the existing infrastructure options also require no or minimal transmission system upgrades to be effected. Load control provision of balancing services is expanding in the USA. Demand response (primarily industrial loads) currently supplies half the contingency reserves in ERCOT. Aluminum smelters have begun to supply regulation to MISO through load control.

No new technological advances are needed to access the storage capabilities inherent in current end-uses. Pilot studies going back at least until 2007 showed the viability of accessing end-use storage (PNNL, 2007). Rheem Corporation in the USA manufactures sophisticated electric water heaters than can respond to grid needs. Steffes Corporation, also in the USA, manufactures both electric space heating storage devices and sophisticated water heater retrofit kits that can operate water heaters beyond their normal temperature range to access greater storage capacity for the purpose of responding to power system balancing needs. Perhaps the greatest challenge is not technological, but in moving from the assumption that generation must respond to load, to a new paradigm where load responds to generation.

14.3 FACILITY SITING

A major challenge relates to siting and permitting wind projects and transmission facilities necessary to transmit energy from typically remote wind-rich regions to more populated areas. Wind generation and transmission facilities are large, industrial-scale structures and present unique challenges for development, particularly in areas with a pre-existing use of the property on which the facilities are sited. A pre-existing use can be just about anything, from wildlife habitat to a scenic area to airspace for military flight paths. Impacts to pre-existing uses can often be avoided, minimized and mitigated, but in certain circumstances difficult tradeoffs will need to be made. For example, the importance of a viewshed to impassioned portions of local populations will need to be weighed against the national and global interests of a secure and relatively clean energy supply in the context of all developable sites.

One approach to addressing these challenges may be a more centralized approach, with established siting criteria and consultation with appropriate stakeholders. In Denmark, again a leader in the industry, government researchers have taken the responsibility for screening the most desirable wind sites from both energy and environmental perspectives. Once the sites have been thoroughly reviewed and analyzed through processes that include local government representatives, the federal government will accept bids from industry for the final development. Responsibility for ensuring sufficient transmission capability to the sites is the responsibility of the national grid operator. In contrast, the USA relies on a patchwork of local, state, and federal regulatory bodies to approve power generation and transmission projects. Wind development companies have fanned out across the country to discover projects to develop. The industry has largely acted responsibly with project developers tending to follow a path of least risk. As the sites with the lowest risk become developed, the tradeoffs between the benefits and impacts of wind development become increasingly challenging to address on a project-by-project basis.

It seems likely that more standardized processes such as seem to be occurring in Denmark will be necessary to achieve wind penetration levels above 50%. There are several efforts in the USA to develop wind energy siting and permitting guidelines, but the lack of a federal standardized regulatory process for project development may represent a major challenge to the development of wind energy as the prime source of electric energy.

14.4 WIND FORECASTING

Efficiently using wind energy requires the best possible wind schedules and forecasts. Some grid operators require wind generation schedulers to participate in region-wide wind forecasting efforts, others rely on the schedulers to supply their own forecasts. Costs and penalties associated with the level of inaccuracy of schedules may or may not accurately reflect the value of the accuracy provided. Without an appropriate economic signal, it has been difficult to prepare economic justifications for improved data collection and processing facilities that would be required to effect improvements in wind schedule accuracy. One of the purposes of this book is to help power system analysts identify the value of generation schedule and forecast accuracy to spur investments necessary to improve forecast accuracy. Certainly, a power system that relies predominantly on wind energy will be a power system that cares much more about the weather than is presently the case in most power systems.

It is likely that the best forecasts will be centrally managed, as opposed to individual wind project schedulers acquiring forecasting services separately for themselves. First, pooling resources across a wider range of participants will make more resources available in general. Second, investments in data collection and processing that benefit a range of system participants are best shared by all the participants and should not alternatively be duplicated by all participants needing the data. Third, improvements in forecasting technology can be implemented for all participants together as opposed to potentially having best practices adopted by some, but not all, forecasters.

To be sure, there are pitfalls in centrally managed systems of any kind. An efficient forecast provider would be one that receives forecasting services from multiple vendors, using the diversity of the forecasts to both obtain the best possible forecast and potentially to gauge the accuracy of the forecasts—widely diverging forecasts may be an indication of uncertainty in the prediction. Another reason to rely on multiple forecast suppliers is to ensure competition and innovation.

At present, measurements of atmospheric conditions (e.g. temperature, pressure, humidity, etc.) as well as important parameters such as soil moisture content and insolation are relatively sparse in time and space. More frequent and more closely spaced measurement taken at various altitudes will be vitally important to improvements in wind energy forecasts. Investments on this scale must either be undertaken at the national level (in the national interest) or by wind generation schedulers or grid operators banding together for the purpose of funding the much higher level of data gathering and analysis.

14.5 CONTROLLING WIND GENERATION

Wind generation exhibits variable and somewhat random behavior. As such, there will be times due to coincidence or coordinated weather systems when extreme events take place. Advanced forecasting techniques may help alleviate some of the effects of rapid coordinated changes in wind generation, but in systems where wind generation accounts for the majority of electric energy generated, extreme events can occur and it may not be economically advantageous to plan a system where even the most extreme events are accommodated using balancing reserves held on more traditional generating technologies.

An alternative is to enact wind generator controls that limit generator levels or rates of increase on generation levels. Such limitations would not be invoked on all hours, but at times when volatile weather events are likely or wind projects across a large region are expected to act in concert, it may be most cost-effective to impose limits. Today limits are largely imposed due to transmission congestion. Such events can occur when strong winds hit an area with a weak connection to distant load centers such as between west and east Texas in the USA. However, the Bonneville Power Administration (BPA) has purposely limited the reserves held for wind facilities on its system, and limits wind generation or delivery schedules when the deployment of reserves nears the amount set aside. This can be a cost-effective strategy, although in the specific case of BPA, limitations are currently imposed regardless of whether other reserve generation is available on the system—not a particularly economically efficient use of the available wind generation.

14.6 SUMMARY

Power systems supplying the majority of energy demand with wind generation will need to shift from a paradigm in which generators generally respond to the dictates of demand, to systems where demand is adjusted to accommodate changes in wind generation. Such systems will make greater use of available storage, primarily in the form of thermal energy, in existing infrastructures such as buildings, water heaters, municipal water pumping systems, and district heating systems where available. Although demand will be controlled by the utilities to some extent, the effects of this control will be almost entirely undetectable by end users.

The present balkanization of many of the balancing areas, mostly in the USA, will need to be addressed so that both balancing needs and balancing services can be shared as widely and economically as possible. Aggregation of balancing areas should foster the more efficient use of transmission capability that can suffer in systems that allow or potentially require reservations.

Getting to a high penetration rate of wind generation will require addressing another type of balkanization, the way in which wind project and transmission facilities are sited. Denmark may provide an example of a more coordinated approach that more directly involves stakeholders in a process that has a higher likelihood of making environmental, aesthetic, and societal tradeoffs than siting processes that occur over a diverse set of local, regional, and national regulatory bodies.

Aggregating wind-forecasting services into regional entities is likely also necessary to facilitate the efficient use of wind generation necessary to achieving high penetration levels. Costs associated with data gathering and processing are most manageable as shared across entities benefitting from the services, and most especially ensuring that data gathering and processing efforts are not unnecessarily duplicated in specific regions.

Finally, it is clear that ever more sophisticated control of wind generation, especially in conjunction with more advanced wind forecasting techniques, has the potential of significantly reducing costs associated with extreme wind ramping events.

REFERENCES

Energinet.dk (2009). *Wind power to combat climate change: how to integrate wind energy into the power system*. Fredericia, Denmark: Energinet.dk. ISBN-13, 978-87-90707-65-1.

Pacific Northwest National Laboratory (PNNL). (2007). *Pacific Northwest GridWise™ Testbed Demonstration Projects: Olympic Penninsula Report*. Prepared for US Department of Energy under Contract DE-AC05–76RL01830. October.

Appendix A: Wind Forecasting Vendors

Commercial wind information service providers offer a variety of services relating to wind generation:

- Wind speed or generation forecasts for the next hour, day, or week.
- Wind resource maps estimating average wind speeds.
- Interpolated (between hours and meteorological stations) historical wind speeds.
- Simulated wind generation over historical time periods based on numerical weather prediction models.
- Site assessment assistance.

Providers can often assist in developing data where the available data are sparse or difficult to interpret.

3Tier Group: www.3tiergroup.com, headquartered in Seattle, Washington, USA with offices in Bangalore, India; Panama City, Panama; and Hobart, Australia.

AWS True Wind: www.awstruewind.com, headquartered in Albany, New York, USA.

Energy and Meteo: www.energymeteo.com, headquartered in Oldenburg, Germany.

Garrad Hassan: http://www.garradhassan.com has offices around the world.

Precision Wind: http://www.precisionwind.com, headquartered in Santa Cruz, California, USA.

WindLogics: www.windlogics.com, headquartered in St Paul, Minnesota, USA.

WSI: www.wsi.com, headquartered in Andover, Massachusetts, USA and Birmingham, UK.

Glossary

When I use a word, it means just what I choose it to mean—neither more nor less.

Humpty Dumpty, with help from Lewis Carroll, author, 1832–1898

ACE: See *Area control error.*

Alternating current: An electrical current that changes magnitude and direction with time. In power systems, alternating current closely follows a sinusoidal pattern with a frequency of either 50 or 60 Hz.

Amp: See *Ampere.*

Ampere (often shortened to 'amp'): A measure of the flow of electricity (current) along some path.

Ancillary services: Services that may be provided by generating units (and sometimes energy-consuming units) that help maintain the reliability of the power system, distinct from the simple delivery or generation of electric energy over recognized operating periods. Such services include regulation, voltage and reactance support, spinning and non-spinning contingency reserves.

Apparent power: The nominal (root mean squared) voltage times the nominal current. When voltage and current are in phase (phase = 0), apparent power is equivalent to the real power. At other phase angles, apparent power is composed of two components, the real power and the reactive power.

Area control error (ACE): A measure of the extent to which a power system's scheduled operations (load, generation, imports, and exports) match the actual operations. It is measured in megawatts and can be computed as often as every few seconds. Regulatory bodies may establish limitations on the allowed size of the ACE. An example of such a requirement might be that the average ACE over each 10-minute period in the month should not exceed some defined limit more than 10% of the time.

Automatic generation control: The automated operation of the power output of designated electric generators within a balancing area to maintain the system frequency standard (60 Hz in the USA, 50 Hz in Europe and Asia) and scheduled interchanges with other balancing areas.

Availability: The percentage of time a generating unit or project is physically capable of producing electric power. Availability is independent of the presence or absence of fuel. For example, a wind turbine could be 100% available in a week in which it did not generate power if the wind speed never reached the cut-in speed during the week.

Available transfer capability (ATC): The amount of transmission capability available for sale over a transmission path after taking account of existing uses and sales of that capability.

Balancing area: The electrically interconnected power system under the balancing area authority's jurisdiction.

Balancing area authority: A transmission service provider with responsibility for ensuring reliability over a defined set of power system components, the balancing area.

Baseload: Refers to electric power generators that are usually operated at maximum, or near maximum, capability on all hours. It is typically used to distinguish between generators that change their output in response to changes in market conditions or overall demand. Baseload generation is typified by low variable costs, rendering output reductions uneconomic in most market conditions. Because wind has low variable costs, some analysts may treat wind as baseload generation in analyses.

Base point: See *Generator set point.*

Busbar: The conductor or group of conductors that serves as a common connection for two or more circuits and is used to interconnect equipment at the same voltage. Wind project busbars are usually taken to be the point at which the balance of the power system electrically connects to the collection of wind turbines that make up the project.

Busbar cost: The cost of a power project prior to adding in any transmission, interconnection, or system integration costs—i.e. the cost of the power plant in isolation from power delivery costs.

Capacitor: A power system component that momentarily stores energy for the purpose of supplying reactance (see *Reactance* and *Phase*). By convention, capacitors are said to supply reactive power. Capacitors are sometimes needed to boost voltages to acceptable levels.

Capacity: One of the most widely used, misunderstood, and many-faceted concepts used in the power industry. The Bonneville Power Administration, a federal power-marketing agency in the USA, cites no fewer than a dozen definitions in its list of official definitions. Capacity is generally a maximum rate of generating, transferring, or consuming energy, measured in either kilowatts or megawatts. The multiplicity of definitions springs partly from the fact that the maximum rate of energy production, consumption, or transfer changes depends on various conditions—including defining exactly what piece or pieces of equipment are providing the capacity, what time of year, for how long, whether unit reliability is included, ambient temperatures, consumer demand, etc. The word 'capacity' will be generally avoided in this book except for 'nameplate capacity value' and 'capacity factor'.

Capacity factor: Expresses the fraction of nameplate generating capability that represents the average output of a power plant. For example, a wind project with 100 MW of nameplate capacity that averages 30 MW over a year (262,800 MWh per year) is said to have a capacity factor of 30%. Capacity factors can be defined over different periods of time.

Capacity value: The maximum incremental peak demand that can be met by an incremental generator unit without adversely affecting power system service reliability—taken to be synonymous with effective load-carrying capability (ELCC).

Central limit theorem: A mathematical theorem stating that the sum of a large number of random processes (with finite means and variances) will approximate a normal distribution.

Chronological economic dispatch models (CEDMs): Computer programs that contain algorithms representing the physical attributes of power systems and economic dispatch decisions made by power system operators. The models simulate power system operations through time in discrete time increments (often each hour).

Concentrating solar power (CSP): Power plants that use lenses or mirrors to concentrate light from the sun into a smaller area for the purpose of more easily collecting the energy at higher temperatures. Higher temperature energy is used to produce steam to power steam turbines for producing electric power.

Contingency: The unplanned and usually sudden failure of a major power system component, usually either a generator or transmission line.

Contingency reserves: Represent an amount of generation that must be able to come online in a relatively short time period (variously within 10 or 30 minutes) to maintain system reliability after a sudden failure of a major power system component (usually a generating unit or transmission line).

Contract for differences: A bilateral financial contract between producers and consumers of electricity that stipulates a fixed price sale of power from a producer to a consumer. The producer sells the specified energy into a liquid power market at prevailing prices, and the consumer purchases the power out of the same market. To the extent the market prices are higher than the stipulated price, the producer reimburses the consumer for the difference. Conversely, if market prices are lower than the stipulated contract price, the consumer reimburses the producer for the difference. Such contracts allow for the equivalent of a bilateral purchase/ sale agreement without the need to reserve transmission capability for the transaction, potentially making more efficient use of available transmission capability.

Control area: See *Balancing area authority.*

Control Performance Standard 2 (CPS 2): A balancing area reliability standard established by the North American Electric Reliability Corporation. CPS 2 requires balancing areas to maintain 10-minute average area control error within established bounds of 90% of all 10-minute periods in each month.

CPS 2: See *Control Performance Standard 2.*

CSP: See *Concentrating solar power.*

Current: The flow of electricity along some path. The level of current is measured in amps.

Cut-in speed: The minimum wind speed at which a wind turbine will begin to produce electric power. Modern wind turbines are typically designed to generate power when the wind speed exceeds 5 m/s.

Cut-out speed: The wind speed above which a wind turbine will cease to generate power. Wind turbines typically are designed to shut down to protect themselves in high wind speeds—typically above 25 m/s.

Decremental reserve: Refers to capability to decrease energy generated within an operating period. See also *Reserves* and *Incremental reserve.*

Demand: The amount of power that would be consumed by end-users of the power system under normative operating parameters (frequency, voltage, etc.) and is often used interchangeably with 'load'. Load may differ from demand if the available generation is insufficient to maintaining operating parameters.

Dispatch: Operating a generating unit or power plant to produce electric power. 'Dispatching' a power plant or unit usually refers to the active management of the generation level of a power plant as needed or desired, in contrast to less controllable power plants such as baseload units.

Dispatchable generators: Generators that can increase or decrease generation levels on command. Generally, these are assumed to be natural-gas-fueled turbines or hydro units. However, coal and even wind power can in some relatively more limited ways be considered dispatchable.

District heating: A system for distributing heat generated in a centralized location for residential and commercial heating requirements such as space and water heating, typically using underground hot water or steam pipes. Heat is increasingly supplied from combined heat and power (also called co-generation) plants that produce both electricity and hot water or steam for other purposes. May also provide cooling in 'district energy' systems.

Economic dispatch model: See *Chronological economic dispatch models (CEDMs).*

Effective load-carrying capability (ELCC): The incremental load that can be met with an incremental generator without diminishing probabilistic reliability measures.

Energy: The ability to do work. Work and energy are defined in terms of a force times a distance. In metric units, force is measured in newtons (about a quarter of a pound) and distance in meters. A 1 newton weight lifted a distance of 1 meter is defined as 1 joule of work; 3600 joules are equivalent to 1 kilowatt-hour.

Equivalent forced outage rate (EFOR): Represents unit failure (unplanned outage hours and equivalent unplanned de-rated hours) as a percentage of the total hours of the availability of a generator. See also *Forced outage rate (FOR).*

Expected unserved energy (EUE): The average (mean) energy demand not met over some specified time interval—often expressed in terms of megawatt-hours per year.

Forced outage rate (FOR): The average percentage of time a generator is unavailable due to unplanned equipment malfunctions. See also *Equivalent forced outage rate (EFOR)*.

Gate closure: The time at which generation and delivery schedules are due for an operating period. Gate closure times are often expressed as a number of minutes prior to the beginning of an operating period.

Gaussian distribution: See *Normal distribution*.

Generator set point: A target electric power generator output level over an operating period. Actual generator output may vary for a variety of reasons, including response to system frequency.

Ground reflectivity: The percentage of sunlight reflecting off the surface of the earth at some point. Low ground reflectivity implies a higher degree of solar energy absorbed by the ground.

Heat rate: The amount of energy consumed by a power plant divided by the energy produced, this represents the (inverse of) efficiency of energy conversion. Heat rates are often expressed in terms of BTU/kWh. Table G.1 shows heat rates in BTU/kWh and the corresponding overall energy conversion efficiency.

TABLE G.1 Equivalent Efficiency Levels at Various Heat Rates

Heat Rate (BTU/kWh)	Conversion Efficiency (%)
3412	100
5000	68
10,000	34
15,000	23
20,000	17
25,000	14

Hertz: A unit of frequency equivalent to cycles per second. For example, North American power grids operate at a frequency of 60 Hz, meaning that the current repeats cycles of reversing direction 60 times each second.

Hub height: The distance between the wind-turbine hub (center of the blades) and the ground. Common hub heights for current generation wind turbines are 80–90 meters.

Incremental reserve: Refers to capability to increase energy generated within an operating period. See also *Reserve Generation* and *Decremental reserve*.

Independent system operator: Coordinates, controls, and monitors the operation of a high-voltage electrical power system. Independent system operators regionalize control of power systems formerly owned and operated by multiple (mostly vertically integrated) utilities.

Inductor (also called 'Reactor'): A reactance device (see *Reactance* and *Phase*) that tends to create a phase difference by causing current to lag changes in voltage. By convention, inductors are said to consume reactive power. Adding inductors tends to reduce voltage levels in the local area.

Insolation: The electromagnetic energy striking the earth's surface, including visible light, originating from the sun.

Integrated power system: Refers to the collection of power generators interconnected together, acting in concert to meet consumer demand in some geographic region. The generators may also respond to markets in adjacent systems to which they may have significant access.

Integrated resource plans: Economic studies resulting in recommendations for resource acquisitions. The methodology is characterized by a holistic approach that examines the overall cost and benefit of added resources to the power systems as a whole. This is distinct from older economic studies of resource additions that focused on the specific economics of the added power plant, potentially without adequate consideration of the economics of the power system as a whole.

Joule: A unit of energy or work, defined as 1 newton (unit of weight) times 1 meter.

Levelized cost: Represents the present value of the total cost of building and operating a generating plant over its economic life, converted to a constant per-MWh value that has the same present value. The operation is mathematically equivalent to dividing the present value of the stream of costs by the discounted stream of energy production. For example, if a generating unit with a 20-year life has a present value cost of $100 million at 5% discount rate and produces 100,000 MWh per year, then its levelized cost is $100 million divided by the discounted energy stream: Levelized cost = $100,000,000/PV (100,000 MWh, 5%, 20 years) = $100,000,000/1,246,221 MWh= $80.24/MWh.

Liquid market points: Specifically designated points on a power system where multiple entities regularly trade significant quantities of power on varying time scales.

Liquidity: The ability to quickly and easily buy or sell into a market without significantly affecting market prices. The liquidity of a market is sometimes considered to be indicated by trading volume (i.e. quantity of trades occurring in a market). Also known as 'market liquidity'.

Load: The consumption of energy at the consumer end of the power system, usually measured as a rate of consumption on an instantaneous basis in megawatts (or kilowatts), or integrated over time as either average megawatts (kilowatts) or megawatt-hours (kilowatt-hours).

Load flow model: See *Power flow model.*

Loaded units: Generators actively generating electric power at some level.

Loss of load expectation (LOLE): The average outage time over some time interval, usually expressed as days, hours, or minutes over a year, 10 years, etc.

Loss of load probability (LOLP): The probability that less than 100% of demand can be met over a given time interval.

Market liquidity: See *Liquidity.*

Merit order dispatch: Operating lower variable (mainly fuel) costs in preference to the higher variable cost resources. At most times, it is not necessary to generate power from every power plant on a power system. Merit order dispatch means (all else being equal) the highest incremental cost resources are the last added to meet load.

Meso-scale models: Computer-based numerical weather prediction models commonly used by meteorologists for forecasting weather. They are commonly employed by wind forecasting service providers as an input to forecasting wind project generation levels.

Minimum generating requirement: The smallest amount of generation a power system can achieve without risking reliability issues. For hydro units, there may be minimum streamflow requirements, potentially imposed for environmental reasons that may be a limiting factor. Larger coal or nuclear-fired generation may be constrained to levels below which the power plants cannot reliably operate.

Nameplate capacity: The maximum rate of energy generation (megawatts or kilowatts) that can be produced by a generating unit or project or project under standardized conditions as stated by the manufacturer.

Negative load: Used to denote analytical methods that assume wind generation is subtracted from load prior to operating the non-wind resources. This is a useful technique at lower penetration levels of wind, but at higher levels the ability to control wind output becomes important and may need to be taken into account.

NERC: See *North American Electric Reliability Corporation.*

Newton: A measurement unit of force or weight. One newton is approximately 0.224 pounds.

Nominal dollars: Economic values expressed in dollar terms in the year for which the expenditure is made. Nominal dollars are unadjusted for the effects of inflation or the time value of money. For example, an expenditure of $100 in 10 years time (nominal dollars) is not worth $100 today. See also *Real dollars*.

Normal distribution: Also known as the Gaussian distribution or bell curve, this represents a particular distribution of numbers around a mean. The distribution is described by the equation:

$$P(x) = \frac{1}{\sigma\sqrt{2\pi}} \exp\left[-\frac{(x-\mu)^2}{2\sigma^2}\right]$$

where $P(x)$ is the probability density at the value x, σ is the standard deviation of the distribution, and μ is the distribution mean.

North American Electric Reliability Corporation (NERC): An international, independent, self-regulatory, not-for-profit organization, whose mission is to ensure the reliability of the bulk power system in North America. See http://www.nerc.com. NERC develops and enforces mandatory reliability standards under the authority of the Federal Energy Regulatory Commission and the Canadian Provinces.

Operating period: The time period over which power is scheduled on a power system. Operating period lengths range from 5 minutes to an hour depending on the rules of the local balancing area authority and its reliability council (in the USA).

Operating reserves: Represents the amount of flexibility a utility maintains, or must maintain, to increase or decrease generation levels to maintain system reliability. There are several sub-categories of operating reserves, such as contingency reserve, regulating reserve, and following reserve. Categories of operating reserve distinguish among the various causes of the need for holding flexibility or the relevant time-scales over which the flexibility may be needed.

Operation and maintenance (O&M) costs: The costs associated with operating and maintaining (as opposed to constructing) a power plant. They are often broken down into fixed and variable components. Variable O&M costs are often treated alongside fuel costs in as much as the expenditures are reduced when the plant is not running.

Peak load-carrying capability: The incremental peak load that can be met by an incremental generating station without reducing the reliability of the power system.

Peaking unit: A generator that has flexibility to start up relatively quickly (a few minutes to an hour or so) to meet spikes in power demand. Peaking units are characterized by relatively low capital costs and high variable (e.g. fuel) costs. They are usually operated from a few percent of the hours of the year up to about 20%, depending on market conditions.

Penetration level: A general term denoting the relative amount of wind on power systems. Several indicators of penetration level are in common use. Penetration level may be expressed as the maximum capability of wind divided by the peak system load. It is also measured in terms of the fraction of energy demand that is met with wind-powered generation. Finally, it can be measured in terms of the maximum wind generation as a fraction of the sum of the minimum generating capability (not all generators can be responsibly reduced to zero output) and maximum export capability. Note that in all cases the penetration level is expressed as a percentage, and the basis (peak demand, energy, or min generation/max export) should be noted.

Petajoules: A measure of energy equal to 10^{15} joules.

Phase: A measure of the extent to which voltage and current move together through time. Efficiency is generally maximized when voltage and current are together (phase = 0).

Photovoltaics: Solid-state devices that convert light energy directly to electric energy.

pJ: See *Petajoules*.

Planning reserve margin (PRM): A traditional indicator of power system adequacy calculated as the difference between nameplate generating capability and the level of peak demand, divided by the peak demand.

Power: The rate at which energy is converted, consumed, or transferred. Instantaneous electrical power is the product of instantaneous current (amperes) times voltage (volts). The basic unit of power is the watt (1 watt = 1 volt × 1 ampere).

Power curves: Charts or tables of numbers that map wind speeds to electrical generation.

Power flow model: Computer simulation models that represent the electrical characteristics of transmission systems to determine how electric power physically moves through power grids. Results of power flow model studies help identify transmission system reliability concerns such as low voltage points and transmission-line overloading. Power flow models come in two broad categories: d.c. power flow models that are less detailed estimates of the movement of power, and the more detailed a.c. power flow models that represent voltages and reactance throughout the system.

Power system: A group of electrically interconnected generators, transmission, distribution, and power-consuming facilities.

Present value: The process of discounting monetary streams to a single value at the start of some time period. The discount rate may include the combined effects of inflation and the time value of money, resulting in a present value expressed in 'real dollars'. Present value can sometimes be thought of as the amount of money that, if saved in a bank today, would be sufficient to cover the stream of payments under consideration.

Primary reserve (primarily European usage): Refers to generating capability immediately responsive and fully available within 30 seconds to power system frequency deviations.

Project: A power-generating station that may be made up of one or more generators supplying power to the grid. Multiple generators in a single project usually share physical structures such as substations and controls.

Public Utilities Regulatory Policy Act (PURPA): Legislation passed by the US Congress and signed into law in 1978 that requires utilities to purchase power from certain qualifying facilities (QFs) at the utilities' avoided costs. Wind generating projects no greater than 80 MW are eligible to become QFs under PURPA.

PURPA: See *Public Utilities Regulatory Policy Act.*

PV: See *Photovoltaics.*

Qualifying facilities (QFs): Generating resources meeting requirements under the 1978 US Public Utilities Regulatory Policy Act (PURPA). Utilities in the USA are obligated under PURPA to purchase the output of such facilities at the utilities' avoided cost. Wind generating facilities of not more than 80 MW are included in the definition of eligible QF resources.

Ramp rate: The rate at which generation or load is changing over time—usually expressed as a number of megawatts per minute. Having sufficient ramp rate capability can be an important factor for power systems with high levels of wind penetration.

Reactance: A property of power system components (generators, loads, and power lines) to cause a phase difference between the current and voltage on a power line. There are two distinct and opposite types of reactance: capacitive and inductive. Reactance is purposely adjusted to reduce the phase difference, and thereby increase the efficiency of power systems.

Reactive power: A component of the apparent power that does not contribute to the effective energy conversion, consumption, or transfer. The reactive component is zero when current and voltage are in phase (phase = 0), and equals apparent power when voltage and current are completely out of phase (phase = ±90°).

Reactor: A power system component contributing inductive reactance to the power system—used synonymously with inductor.

Real dollars: Economic value expressed in terms of some basis year dollar value, adjusting nominal dollar values for inflation to arrive at a consistent basis for comparison. For example, a payment of $100 made 10 years from now on (i.e. nominal dollars 10 years from today) might be equivalent to $65 today (real dollars, using the current year as a basis) when inflation and the time value of money are considered. See also *Nominal dollars*.

Real power: The effective rate of energy conversion, consumption, or transfer, usually synonymous with power. Real power equals apparent power when the phase is zero, and falls to zero when current and voltage are completely out of phase (phase = $\pm 90°$).

Regulating reserve: A category of operating reserve that utilities hold to maintain power system reliability on a timescale of a few seconds to a few minutes. Regulating reserve is usually held on generating units that can quickly change their output levels (up or down) based on an automated signal (see *Automatic generation control*). Regulating reserve may also be provided by storage devices and from responsive loads.

Renewable energy certificates: See *Renewable energy credits*.

Renewable energy credits (RECs; also tradable renewable certificates (TRCs), or green tags): Certificates of renewable energy production, conveying ownership rights to the environmental attributes of the energy generated. Usually, one REC certifies that 1 megawatt-hour of renewable generation has occurred at a specified generating project within a specified time window (typically the year or quarter in which the energy was generated). Certificates may be traded separately from the actual energy deliveries. RECs are sometimes purchased by utilities to satisfy renewable generation standards enacted by governmental bodies. Individuals or institutions may purchase RECs to enhance the market for renewable energy, and claim the legal right to represent that their energy consumption (or some fraction of it) is met by renewable energy.

Renewable portfolio standards (RPS): Minimum target levels of new renewable energy generation required by some state governments in the USA. Also sometimes called 'renewable energy standards'.

Reserve generation: Generating capability available at some time to respond to unexpected events in order to maintain power system reliability. See also *Contingency reserve, Following reserve, Imbalance reserve, Planning reserve, Primary reserve, Regulating reserve, Secondary reserve, Spinning reserve,* and *Tertiary reserve*.

Reserve requirement: The amount of aggregate reserve generation deemed necessary to maintain power system reliability. See also *Contingency reserve, Following reserve, Imbalance reserve, Planning reserve, Primary reserve, Regulating reserve, Secondary reserve, Spinning reserve,* and *Tertiary reserve*.

Run-of-river hydro: Refers to hydropower projects that have no significant amounts of reservoir storage and are relegated to passing inflows. The nature of run-of-river hydro projects is similar to wind—a relatively variable and less predictable and controllable resource. A significant difference is that generally inflows to hydro projects are less variable than wind due to the natural storage effects of large water collection basins (watersheds) over which the fuel is collected, and the storage that exists in soils and the action of flora.

Schedule error: The difference between the planned (scheduled) generation or load and the generation or load that occurs during an operating period. Schedule error may be taken as an instantaneous value, averaged over time periods as long as the operating period (the latter also called 'schedule imbalance'). Reserve generating units respond to the net schedule error on a power system.

Schedule imbalance: The operating period average difference between the scheduled generation and the actual generator output.

Schedules: Nominated quantities of energy to be generated or delivered over an operating period. The actual energy generated or delivered over the operating period may be different than the levels set in the schedule.

Solar influx: The amount of solar energy incident to the earth.

Spinning reserve: The amount of generation that can respond immediately to increase or decrease output levels. Typically (but not exclusively) spinning reserve is provided by generators physically spinning and synchronized to the power grid, but are operating at less than full output.

Station service: The provision of electric power to a generating station by the power grid. Many generating stations require a power source to begin generating power. Power is used by various pumps and control equipment. Wind turbines also use station service to turn the turbine blades into the wind and adjust turbine blade pitch.

Stochastic analysis: A method of studying random processes by choosing pseudo-random numbers from probability distributions representing the individual random processes.

Stochastic model: A computer model that simulates random processes by choosing pseudo-random numbers from probability distributions representative of random processes such as generator availability levels or wind speeds.

Streamflow: The amount of water that would come down a river system without reservoir operations that alter the actual flows in the rivers. Extremes of streamflow conditions occur during climatic droughts, high precipitation events, or faster than normal melting of mountain snow packs.

Thermal unit: An electric power generator powered by heat energy, either through a steam turbine (as in natural gas, coal, geothermal, or nuclear power plants) or gas turbines (jet engine technology).

Time shift: Refers to the analytical method in which the behavior of a wind project is estimated from the behavior of another up- or downwind project by assuming that what happens at one project is replicated in the second project at some earlier or later time. This technique has sometimes been used to represent the output of a prospective wind project from data collected by an existing one.

Unit: A single generator or generator/turbine couple within a generating station.

Unit commitment: Refers to the logic employed by dispatch models or operators to decide when it is appropriate to start up a generator that is idle, or to shut down an operating generator. Such decisions are usually predicated on the expected economic conditions (wholesale electric market prices, fuel costs, and power plant efficiency) and individual power plant characteristics (startup time, shutdown time, minimum down time, maximum starts per year) while ensuring generation adequacy to meet demand.

Vertically integrated utility: A power company owning most or all of the generating equipment, high-voltage transmission lines, and lower voltage distribution lines to take power from the point of generation to the point of delivery to the ultimate end-user of the power. One of the goals of deregulation is usually to diversify the ownership of generation, transmission, and distribution functions of vertically integrated utilities.

Volt: A unit of electrical potential difference between two points. Often the earth is used as one of the two reference points. Electrical potential difference is effectively an electrical pressu[illegible] cause an electrical current to flow.

V[illegible] al difference or electrical pressure between two points. [illegible] n units of volts.

W[illegible] nd speeds or energy generation that occur due to upwind [illegible] d impinging on downwind turbines.

W[illegible] undary between two air masses in the atmosphere with [illegible] ies. Passage of weather fronts is usually accompanied by [illegible] s.

We[illegible] **Council (WECC):** The regional entity responsible for [illegible] g bulk electric system reliability in western North [illegible]

Win[illegible] represents average wind speeds at a site (see *Table G.2*).

TABLE G.2

Wind Power Class	Wind Speed at 10 Meters (m/s)
1	<4.4
2	4.4–5.1
3	5.1–5.6
4	5.6–6.0
5	6.0–6.4
6	6.4–7.0
7	>7.0

Wind shadow: Refers to the reduction in wind velocity that occurs downwind of a wind turbine. Turbines placed in close proximity to one another in the direction of the prevailing wind can result in a significant decrease of resource for the downwind turbines.

Wind shear: The change of wind speed or direction with elevation at a specific location and time. Low wind shear suggests a condition in which wind speed and direction vary little with height above ground. Conversely, high wind shear implies significant changes in wind speed or direction with height.

Wind turbulence intensity: A measure of the short-term (seconds to minutes) variability of wind speed.

Index